Applied Mathematical Sciences
Volume 131

Springer
New York
Berlin
Heidelberg
Barcelona
Budapest
Hong Kong
London
Milan
Paris
Santa Clara
Singapore
Tokyo

Applied Mathematical Sciences

(continued following index)

J.L. Ericksen

Introduction to the Thermodynamics of Solids

Revised Edition

With 25 Illustrations

 Springer

J.L. Ericksen
5378 Buckskin Bob Drive
Florence, OR 97439-8320
USA

Editors
J.E. Marsden
Control and Dynamical Systems, 116-81
California Institute of Technology
Pasadena, CA 91125
USA

L. Sirovich
Division of Applied Mathematics
Brown University
Providence, RI 02912
USA

Mathematics Subject Classification (1991): 73B30, 7301, 76A15, 80A20

Library of Congress Cataloging-in-Publication Data
Ericksen, J. L. (Jerald L.), 1924–
    Introduction to the thermodynamics of solids / Jerry L. Ericksen.
  — Rev. ed.
    p.   cm. — (Applied mathematical sciences ; 131)
    Includes bibliographical references and index.
    ISBN 0-387-98364-3 (hardcover : alk. paper)
    1. Materials.   2. Thermodynamics.   I. Title.   II. Series: Applied
mathematical sciences (Springer-Verlag New York Inc.) ; v. 131.
    TA403.6.E74   1997
    621.402′1—dc21   97-37938

Printed on acid-free paper.

The first edition of this book was published by Chapman & Hall, UK, © 1991.

Production managed by Steven Pisano; manufacturing supervised by Jacqui Ashri.
Photocomposed pages prepared from the author's LaTeX files.
Printed and bound by Maple-Vail Book Manufacturing Group, York, PA.
Printed in the United States of America.

9 8 7 6 5 4 3 2 1 (revised edition)

ISBN 0-387-98364-3 Springer-Verlag New York Berlin Heidelberg SPIN 10647553

To my dear wife and friend,
Marion E. Ericksen

Preface

Although I have never been actively engaged in research on thermodynamics, *per se*, I have had to utilize it while attempting to develop theories better able to deal with various kinds of nonlinear macroscopic phenomena encountered in materials. In trying to help younger workers to start in research of this kind, I have been impressed with how little most of them know about common and elmentary applications of thermodynamics to solids, although they have taken at least one elementary course in thermodynamics at some university. Observation of a number of elementary books on thermodynamics indicates that this is due more to lack of exposure to such ideas than to some fault of the students.

When my department had a service course become obsolete, I accepted responsibility for developing a replacement course dealing with such applications. The intended audience consisted of seniors and beginning graduate students from various engineering and scientific departments. To make the course accessible to the various groups meant keeping the prerequisites to a minimum, so I settled on mastery of calculus as the basic requirement.

The first nine chapters of this book represent lecture notes developed for this purpose. In the actual lectures some constituents of the tenth chapter are mentioned, but this has not been an integral part of the course. Additionally, a number of simple demonstration experiments are used to illustrate, in a rather crude way, the real phenomena that are being analyzed, but these are not described here.

These notes have also been used for self-study by persons more adept in mathematics and mechanics. For example, I advise graduate students in these disciplines to save time by doing this instead of taking the course. In this way, the course has helped to fill in the gap in education mentioned at the beginning. Chapter 10 is part of the package for such readers. This covers some of the difficulties encountered in trying to apply thermodynamics to obtain a better understanding of the phenomena encountered in solids, partly to indicate the need to grasp the basic concepts of classical thermodynamics.

Unfortunately, different experts in thermodynamics have discordant ideas as to what these basic concepts are. In as elementary a manner as possible, therefore, I discuss what they are as I understand them. However, this chapter is not so elementary as it deals with matters which are unsettled and controversial. In this treatment, classical thermodynamics is interpreted to exclude an important branch based on molecular theory, that is, statistical thermodynamics, only because covering this in any reasonable way would make the notes excessively long. The intent is to provide a small bridge to newer work in thermodynamics.

I have tried to choose a few references which seem likely to be useful, rather than attempting to include all that may be of interest. Since the notes have been used by readers from quite varied backgrounds, references cited range from the very elementary to rather sophisticated works. Readers will need to pick a subselection.

I do not intend this to be a replacement for other elementary books on thermodynamics and do assume that the reader has a little familiarity with the subject. For those who feel a need for supplementary reading, I note a few of the many possibilities. I do think it desirable for writers of elementary books to better cover applications to solids.

Historically, the subject first emerged from studies of ancient heat engines, with a corresponding emphasis on dynamic processes. In particular, this produced early ideas about energy and entropy as they are related to the old laws of thermodynamics. Much in this spirit, although more modern in style, is the book by Truesdell and Bharatha [1]. What has become a more conventional view is that energy and entropy are related more to equilibrium states although one may be dealing with nonequilibrium processes. For a development of the subject from this point of view, the reader is referred to the work by Kestin [2], for example. He is influenced by Gibbs's ideas concerning equilibrium theory, which will be discussed here.

One of the most ardent proponents of the notion that energy and entropy must be related to equilibrium states is Tisza [3], although he is very critical of Gibbs's ideas. It is not clear how many really believe that this view is a "law" of thermodynamics, but most writers of elementary books abide by it. Included are some who exhibit independence of thought in considering various other ideas about thermodynamics. In this category is Pippard, whose book [4] is interesting for its comments and examples, including some relating to solids.

Others, myself included, see no grounds for accepting the restriction and some reason to consider alternatives. In the first nine chapters of this book, the theories treated do fit this mold fairly well, with some caveats mentioned in Chapter 10. However, in Chapter 2, I introduce ideas which are now being used frequently by those willing to consider alternatives. This is one way of gaining experience with the Clausius–Duhem inequality which is accepted and used in studies of irreversible processes by some who, in practice, do accept the conventional view mentioned above, Kestin in particular. Many elementary books do not mention this old inequality.

Rather obviously, it is probable that theories of equilibrium will fit the conventional mold and it would have been easy to include many more examples of this kind. For example, the basic ideas are made available to take the three-dimensional linear theory of thermoelasticity and deduce all the useful inequalities satisfied by moduli using the Clausius–Duhem inequality and the thermodynamic theory of stability. My experience is that this is not a matter of common knowledge among experts in this area, but applications of this kind are fairly routine and useful.

It is also very easy to find theories of solids that are commonly used, some quite old, which do not fit the conventional mold, as was mentioned earlier. It can then be very difficult to know how best to try to apply ideas of thermodynamics. One who has wrestled hard with problems of this kind is apt to see the subject in a somewhat different light. In this category is Bridgman [5], who had a strong interest in applications to plasticity, in particular, although he had no great success in mastering the difficulties involved. Readers unfamiliar with elementary continuum mechanics may find helpful the text by Bowen [6].

Particularly in discussions with chemists I have encountered another prejudice. Roughly, it is that if a material really attains equilibrium, any shear stress must have relaxed to zero. Perhaps this is why so many authors of works on thermodynamics consider only problems of solids subject solely to a hydrostatic pressure, if at all. In this respect most of the examples to be considered illustrate rather common practices of users which tend to be ignored by authors of texts. Considering how the various structures we make tend to deteriorate, I do concede that those chemists have a point.

If we grant it, equilibrium theory for solids should be similar to that for fluids. Then the different theories of thermoelasticity which appear to be equilibrium theories are not really of this kind, despite appearances. If I accepted this and Tisza's view, as I interpret it, thermoelasticity theory is based on improper usage of thermodynamics. I think that it may well be that there is something deeper to be understood here, which could well influence and improve our understanding of energy and entropy. In Chapter 10, I will say a little more about this. However, it often happens that theories which prove to be successful were arrived at by infirm reasoning and this alone is no reason to reject them. I subscribe to the view that, if we can understand more fully why such theories succeed, we will improve our chances of constructing a still better theory.

So, almost everything to be discussed is, to some degree, controversial. I do not enjoy controversy. However, it seems hard to avoid it when, as here, the various experts seem unable to come to agreement as to what should be considered to be the basic concepts and laws of the subject.

In terms of basic content, this revised edition differs from the first edition by including lists of exercises. I have added some comments that did not come to mind when I prepared the first edition and a few references to provide better coverage of some topics. Otherwise, there are some minor corrections and numerous changes in wording.

Last, but not least, I wish to thank those who have assisted me in this enterprise. Some of these have helped me to clarify my own views in debates over matters of principle, particularly Bernard Coleman, Joseph Kestin, Ingo Müller, Paul Naghdi, Ronald Rivlin, James Serrin and Clifford Truesdell. Alan Gent produced very helpful comments concerning my coverage of elastomers and adhesion. A careful reading of early drafts by Millard Beatty and Antonio DeSimone enabled me to eliminate numerous misprints and other slips. As a person who dislikes writing, it is unlikely that I would have brought this to completion without the encouragement of Gunhard Æ Oravas and Patarasp Sethna. Finally, I thank Kathryn Kosiak and Lee Reynolds for much help in preparing the manuscript.

J.L. Ericksen

Contents

1
Generalities

1.1 Energy, Heat, and Power

Thermodynamic systems are capable of occupying a variety of states linked by time-dependent processes. Often we think of a process as leaving one state and arriving at another or passing through a succession of states. To induce a system to undergo the processes of which it is capable, we generally need to bring it into contact with, or let it interact with, various other kinds of systems.

The variables needed to describe states properly are different for different systems and, for a given system, can depend on the range of situations to be considered. For solids, some measures of strain and temperature are likely to be included. Such things as loading devices of different kinds, thermostats, and so on, will be used to induce changes of such states and need to be described. If one is concerned with, say, liquid crystals, one needs also to introduce appropriate variables describing orientation and electromagnetic fields, as will be discussed in Chapter 9. Similar remarks apply to the description of processes.

For a thermodynamic system, the first law asserts that, for any possible process,

$$\frac{dE}{dt} = P + Q, \qquad (1.1.1)$$

where E is the energy of the system, P is the power, that is, the rate at which work is done on the system, and Q is the rate at which heat is supplied to the system. Intuitively, other systems do supply the power and heat. However, there is some implication that, in a theory for a system, we should have formats for calculating P, Q, and E for any possible process.

Often, we consider a system to be in contact with another system that interacts with it in a special way. In terms of how P may be affected, some of the possibilities of interest include the following:

$$P = 0 \quad (mechanically\ isolated\ system). \tag{1.1.2}$$

This can occur because no forces are applied or, often, because forces are applied but motions are restricted by rigid walls, and so on so that the forces do no work. For systems we think of as power sources, it is reasonable to assume that

$$P \leq 0 \quad (mechanically\ passive\ system). \tag{1.1.3}$$

Systems loaded by gravity, springs, and so on, often fit the description

$$P = -d\chi/dt \quad (conservative\ loading\ devices). \tag{1.1.4}$$

Here, it is usually understood that χ can be calculated given the relevant state variables. Various little sources of dissipation in a loading device, or adding damping mechanisms, can shift (1.1.4) to

$$P \leq -d\chi/dt \quad (dissipative\ loading\ device). \tag{1.1.5}$$

In terms of Q, we could have, with suitable insulation,

$$Q = 0 \quad (thermally\ isolated\ system). \tag{1.1.6}$$

Systems designed to give off, but not accept, heat fit

$$Q \leq 0 \quad (thermally\ passive\ system). \tag{1.1.7}$$

If a system is not in contact with any other system, we expect that

$$P = Q = 0 \Rightarrow E = \text{const.} \quad (isolated\ system). \tag{1.1.8}$$

If the environment is such that either (1.1.6) or (1.1.7) and one of (1.1.2)–(1.1.5) applies, the intuitive expectation is that the system should approach equilibrium. However, at this stage, it is not easy to assign any reasonably clear meaning to the statement.

A pattern of thought underlies measurements of energies for thermodynamic systems. Consider any pair of states labelled 1 and 2, a process leaving 1 at time t_1 and arriving at 2 at time t_2; then, integrating (1.1.1) gives

$$E_2 - E_1 = \delta W + \delta Q, \tag{1.1.9}$$

where

$$\delta W = \int_{t_1}^{t_2} P\, dt, \qquad \delta Q = \int_{t_1}^{t_2} Q\, dt \tag{1.1.10}$$

represent the total work done on the system in the process and the total heat supplied, respectively. There may be many processes connecting the two states giving different values of δW and δQ but, as the first law is commonly interpreted, the value of $E_2 - E_1$ should be the same for all; it depends only on the end states. For one of the possibilities, experimental data is required enabling the estimation of δW and δQ. Calorimetry provides methods for measuring δQ, rather routine in some cases. Measuring δQ for a solid held in a loading device is not so easy and it certainly helps to be able to think of 1 and 2 as being static equilibrium states. To some degree, estimating work from measurements of forces and displacements is a familiar problem in mechanics. However, because of the complicated methods commonly used to grip solids, in practice it can be very hard, perhaps impossible, to make an accurate estimate of δW. Consider 1 as a fixed state and 2 as any other state. Then, assuming one can get the requisite experimental estimates, E can be determined, to within a constant, for each of the possible states. Physically, only energy differences have significance, so one can fix the constant in any convenient way.

We have discussed two rather different views of energy. In (1.1.1), it is a function of t, somehow calculable for any possible process. Then, in the later discussion, it becomes associated with states. To combine the two, one needs to make an assumption about the relation between states and processes—basically that, at each point in time, a process uniquely determines a state. Then, for a process, $E(t)$ is the value of E for the associated state.

1.2   Temperature and Entropy

Entropy and temperature play important roles in thermodynamics. The reader is assumed to have some familiarity with the notion of absolute temperature, which we denote by θ, with values satisfying $\theta > 0$. In terms of this, we can describe another kind of thermal environment, that of an ideal heat bath, whose temperature

$$\theta = \theta_B(t) \tag{1.2.1}$$

depends possibly on time t but not on position. Consistent with this is the idea that putting a system in contact with the heat bath does not change $\theta_B(t)$, although the temperature in the system can differ considerably from this value. Think, as Newton did, of a small hot poker cooling off in a large room. It will be hot to the touch, but θ_B, interpreted as room temperature far from the rod, will not change much. In other systems, we employ cooling devices, and so on, to keep the temperature of at least part of a system near a definite value, identified as θ_B. Commonly, changes in entropy are inferred from measurements in an environment of this kind, with θ_B held constant for quite a while, changed a little, then held at a new constant value.

Such determinations of entropy involve an assumption which is rather restrictive for solids: every pair of states can be connected by at least one reversible process, or at least by approximating this situation. Physically, this tends to exclude inherently irreversible phenomena such as are encountered in sliding friction or plasticity; so we will ignore such exceptions. The idea is that the entropy S of a system in contact with a heat bath should satisfy what is often called the *Clausius–Planck inequality*

$$\frac{dS}{dt} \geq \frac{Q}{\theta_B} \tag{1.2.2}$$

in the processes that are possible. Here, there are no restrictions as to how power can be supplied to the system: it is only the thermal environment that is somewhat special. Involved is the idea that, in (1.2.2), equality holds for reversible processes.

As in the discussion of energy, we think of processes starting at state 1 at time t_1 and arriving at state 2 at time t_2. Then, integrating (1.2.2) gives, for the entropy difference,

$$S_2 - S_1 \geq \int_{t_1}^{t_2} (Q/\theta_B)dt. \tag{1.2.3}$$

Next, we try to find a reversible process, reducing (1.2.3) to

$$S_2 - S_1 = \int_{t_1}^{t_2} (Q/\theta_B)dt. \tag{1.2.4}$$

Since Q is not easy to measure, the usual practive is to try to arrange that θ_B is constant, so that we have

$$S_2 - S_1 = \delta Q/\theta_B, \tag{1.2.5}$$

where δQ is given by (1.1.10). Then, of course, δQ may be measured by calorimetric techniques. With θ_B also measured, we get experimental values of the entropy difference between two states. We do this for as many states as is feasible. In principle, this determines values for S for such states to within an unimportant additive constant.

From what has been said here, it is unclear as to what properties entropy may have or how it is to be defined or measured when the environment does not fit the above "heat bath" prescription. Common practice involves, at least tacitly, the assumption that from measurements in such environments one can infer values of S for states occurring in more general kinds of processes. One should appreciate that "common" has a different meaning from "universal." Said differently, an intelligent person does not accept popular practices thoughtlessly.

In a famous paper on thermodynamic equilibrium, Gibbs [7] began with a quotation from Clausius:

Die Energie der Welt ist constant.
Die Entropie der Welt strebt einem Maximum zu.

It is hard to be sure exactly what Clausius meant by "der Welt," but later writers, including Gibbs, interpreted that as meaning an isolated system. Then, a translation is:

The energy of an isolated system is constant.
The entropy of an isolated system tends to a maximum.

Clearly, the first statement agrees with (1.1.8). Related to the second statement is a widely accepted idea (concerning isolated sytems) that

$$P = Q = 0 \Rightarrow dS/dt \geq 0. \tag{1.2.6}$$

In fact, it is generally accepted that the restriction to mechanically isolated systems is not necessary, although that to thermally isolated systems is. That is, (1.2.6) can be replaced by

$$Q = 0 \Rightarrow dS/dt \geq 0. \tag{1.2.7}$$

It is hard to compare either of these statements with (1.2.2), since it is not reasonable to think of a thermally insulated system as being in contact with a heat bath.

As to what energy and entropy meant to Gibbs, we have his rather terse statement:

As the difference of the values of the energy for any two states represents the combined amount of work and heat received or yielded by the system when it is brought from one state to the other, and the difference of entropy is the limit of all of the possible values of the integral $\int \frac{dQ}{t}$ (dQ denoting the element of heat received from external sources, and t the temperature of the part of the system receiving it), the varying values of the energy and entropy characterize, in all that is essential, the effects producible by the system in passing from one state to another.

There is a rather common view that no real process is quite reversible, so reversible processes may be thought of as limits of processes which are not themselves real processes. Perhaps Gibbs's use of the phrase "the limit" indicates that he held this view. Some tend to associate nearly reversible processes with those taking place very slowly—the "quasi-static" processes. This probably stems, at least in part, from experience with classical theories of heat conduction, viscosity, and so on. However, a ball made of silly putty behaves almost reversibly when bounced rapidly and various other high polymers have similar predilections. So, it seems prudent to be open-minded in considering what may be reversible processes for particular systems.

It is a blunt fact that the general ideas of entropy and, to some extent, power, heat, and energy remain somewhat nebulous. Despite this, they have been used quite successfully in analyzing various kinds of physical systems. In the first eight chapters of this book the aim is modest. It is to discuss some of the kinds of thermodynamic ideas commonly used in analyzing the behavior of solids, illustrating this by relatively simple examples. Chapter 9 serves to provide a broader perspective. After acquiring this experience, we will reconsider the basic concepts in Chapter 10.

1.3   Thermodynamic Equilibrium

Of considerable importance are general ideas of equilibrium and the related stability theory. Intuitively, systems can be expected to approach equilibrium only if they are kept in rather special environments. In particular, we expect a trend to equilibrium in physically isolated systems. Gibbs took this intuitive idea together with older thermodynamic ideas discussed in Section 1.2 and used them to motivate definitions of thermodynamic equilibrium, and criteria for the stability thereof, for isolated systems. In his words,

> I. *For the equilibrium of any isolated system, it is necessary and sufficient that in all possible variations of the state of the system which do not alter its energy, the variation of its entropy shall either vanish or be negative.*

He also gave a plausible argument[1] indicating that this statement is equivalent to

> II. *For the equilibrium of any isolated system, it is necessary and sufficient that in all possible variations in the state of the system which do not alter its entropy, the variation of its energy shall either vanish or be positive.*

Discussion following this indicates that one should exercise some good physical judgement in deciding as to what are the "possible variations." This kind of difficulty occurs in all forms of stability theory for rather simple reasons. Nothing, it seems, is indestructible, so it is too much to ask that a system remain stable with respect to every conceivable kind of disturbance. Some quite unstable systems, such as a room filled with natural gas, can tolerate some small vibrations but be set off by a small spark. Basically,

---

[1] It is hard to see how one could give a general proof of this since the statements are not very precise, serving more as guidelines. When there is any doubt about equivalence, workers take as basic the first statement.

we want to know whether a particular system will be stable with respect to those disturbances that are likely to be encountered but are beyond our control. So, inevitably, we must exercise some judgement in deciding what these might be. As was mentioned before, when one encounters phenomena such as sliding friction or plasticity, entropy is ill-defined so that statements are not applicable. Curiously, Gibbs did not mention this difficulty but did warn against trying to use the criteria in such cases. His is a long memoir, discussing many types of physical problems fitting this mould and including such topics as chemical reactions in fluids, absorption of fluids by solids, dissolving solids in fluids, stability of fluids surrounded by solids and vice versa, as in pressure vessels, and so on. It is not easy to read but the density of ideas is high, and many are now commonly used by engineers and scientists of various disciplines.

Often, the use of thermostats makes it more natural to think of a system not as thermally insulated but as in contact with a heat bath at a constant temperature θ_B. Then, from (1.1.1), with the system considered as mechanically isolated, we have[2]

$$\frac{dE}{dt} = Q \qquad (1.3.1)$$

and, from (1.2.2),

$$Q \le \theta_B \frac{dS}{dt},$$

so

$$\frac{dE_B}{dt} \le 0. \qquad (1.3.2)$$

where

$$E_B = E - \theta_B S. \qquad (1.3.3)$$

Physically, one expects an approach to equilibrium in such systems, with a caveat. There exists the phenomenon of Brownian motion which is rather easy to observe in fluids. Statistical theory[3] leads us to expect such motion to occur in solids, little fluctuations preventing the system from quite reaching equilibrium. Usually, the motion is so small that errors that result from neglecting it is of little practical significance and we will ignore it. Then, with E_B decreasing in any process, we expect that E_B will be some kind of a minimum in stable equilbrium. The quantity E_B is called the *ballistic free energy*. This suggests a third statement which we write, in the style of Gibbs, as

III. *For the equilibrium of any mechanically isolated system, in contact with a heat bath at constant temperature, it is*

---

[2]In this discussion, we lean rather heavily on ideas of Duhem [8].

[3]Early work on this, performed by Einstein, is collected in reference [9].

> *necessary and sufficient that in all possible variations of the system the variation of its ballistic free energy shall either vanish or be positive.*

As will become clear, this gets us rather close to the "energy criteria" commonly used by engineers to analyze stability of various structures.

In any of these situations, the systems could contain fluids of unknown viscosity—various other kinds of dissipative mechanisms which we might not know how to describe quantitatively. More often than not this is the situation encountered in practice. From the statements it is clear that one does need prescriptions for calculating energies and entropies. Given this, one can use these ideas to find and test the stability of equilibria. It is not a trivial matter to find good prescriptions. Later we will discuss some rather common kinds of difficulties that do arise in practice. Also, we will encounter cases where predictions obtained depend on the selection of the "possible variations" among choices which seem rather likely.

Common to the three statements is the restriction that the system be mechanically isolated. By lumping together solid specimens and devices used to load them one can meet this requirement, in a reasonable way, in various situations of physical interest. It is common to relax this requirement somewhat by admitting loading devices that are conservative or dissipative, in the sense indicated by (1.1.4) or (1.1.5), respectively, then regarding the solid specimen as the thermodynamic system. For example, we might load a specimen by attaching a bucket, to which we will add fluid, in the earth's gravitational field, identifying χ with the gravitational potential. In itself the gravitational force is conservative, but any motion of the fluid could induce viscous dissipation and (1.1.5) can cover this. By an obvious modification of the reasoning leading to (1.3.2), one now obtains

$$\frac{d}{dt}E_\chi \leq 0, \tag{1.3.4}$$

where

$$E_\chi = E + \chi - \theta_B S, \tag{1.3.5}$$

which we shall also call the *ballistic free energy* of such systems. This serves to motivate a fourth statement:

> IV. *For the equilibrium of any system in a conservative or dissipative loading device, in contact with a heat bath at constant temperature, it is necessary and sufficient that in all possible variations of the system the variation of its ballistic free energy shall either vanish or be positive.*

In practice, workers often use III or IV, simplified by the assumption that, throughout the system, the temperature is θ_B, although the "possible variations" generally include other kinds of temperature distributions. Such criteria are slightly inferior in principle, but the differences tend to be compensated for by other assumptions accepted by such workers, sometimes tacitly. Later, we will elaborate on this for special cases.

The four statements summarize ideas that have been used with considerable success to analyze stability of equilibrium for a great variety of physical systems. One could generalize to cover passive systems, but, for arbitrary thermodynamic systems, thermodynamic criteria for stability do not exist and there are the aforementioned difficulties in covering such topics as plasticity or Brownian motion. So, the thermodynamic theory of stability has its limits. Other approaches to stability theory are covered, in a rather elementary way, in Pippard [10], for example.

2
Constitutive Theory of Heat Transfer for Bars and Plates

2.1 Thermodynamics of Rigid Bars

In this section, we consider the one-dimensional theory of heat transfer in rigid, stationary bars, illustrating how some of the general thermodynamic ideas are interpreted in this context. Mathematically, the points on a bar are represented by points in the interval

$$0 \leq x \leq L. \tag{2.1.1}$$

Quantities of interest, such as temperature, will then, in any process, be functions of x and t. We require a theory capable of describing the possible temperature distributions. One rather general idea is that a part of a thermodynamic system can itself be considered as a thermodynamic system, so, in particular, any part of the bar qualifies. In Chapter 8, we will note that it is not always feasible to use this idea. Forces may act on any such part but these will do no work because motion is excluded. Thus, any such part, or the whole bar, will be mechanically isolated,

$$P = 0. \tag{2.1.2}$$

Another general idea is that energy and entropy are additive.[1] Here, the traditional way of covering this is to assume that they are representable by

---

[1] Generally, thermodynamicists call such quantities extensive variables, referring to quantities which are not, for example θ, as intensive variables.

integrals

$$E = \int \varepsilon \, dx, \qquad S = \int \eta \, dx, \qquad (2.1.3)$$

where the integral can be over the interval (2.1.1), or a subset.

Physically, heat transfer can take place along the bar by conduction. To describe this, we introduce a heat flux q. In terms of this Q_c, the rate at which heat is supplied by this mechanism to a subinterval (x_1, x_2) is represented by

$$Q_c = q(x_2, t) - q(x_1, t) = q\Big|_{x_1}^{x_2}. \qquad (2.1.4)$$

With one-dimensional theory, it is difficult to describe transfer through the sides of the bar in a similar way, so we introduce another representation,

$$Q_b = \int_{x_1}^{x_2} r \, dx, \qquad (2.1.5)$$

sometimes considered as a measure of radiation, sometimes as a crude approximation to the effects of conduction or perhaps some combination of both. Then, for the subinterval, (1.1.1) takes the form

$$\frac{dE}{dt} = \frac{d}{dt} \int_{x_1}^{x_2} \varepsilon \, dx = Q$$

$$= Q_c + Q_b \qquad (2.1.6)$$

$$= q(x_2, t) - q(x_1, t) + \int_{x_1}^{x_2} r \, dx.$$

Assuming the functions are smooth enough, we can rewrite this as

$$\int_{x_1}^{x_2} \frac{\partial \varepsilon}{\partial t} \, dx = \int_{x_1}^{x_2} \left(\frac{\partial q}{\partial x} + r \right) dx, \qquad (2.1.7)$$

and infer from the arbitrariness of x_1 and x_2 that

$$\frac{\partial \varepsilon}{\partial t} = \frac{\partial q}{\partial x} + r. \qquad (2.1.8)$$

In this context, the accepted view is that the entropy is involved in an inequality, the so-called *Clausius–Duhem inequality*,

$$\frac{d}{dt} \int_{x_1}^{x_2} \eta \, dx \geq (q/\theta)\Big|_{x_1}^{x_2} + \int_{x_1}^{x_2} (r/\theta) \, dx, \qquad (2.1.9)$$

reducing to the local form,

$$\frac{\partial \eta}{\partial t} \geq \frac{\partial}{\partial x}(q/\theta) + r/\theta. \qquad (2.1.10)$$

Additional equations are needed, and physically these should take into account that the thermal response will be different for different materials.

Before considering this, let us consider some of the different ways of describing the notion that the bar is in contact with a heat bath at temperature $\theta_B(t)$, a given function of t. One possibility, envisaging that the side of the bar is coated with an insulator, is

$$r = 0, \qquad \theta(0, t) = \theta(L, t) = \theta_B(t). \tag{2.1.11}$$

Then, applying (2.1.9) to the whole bar, we get

$$\frac{dS}{dt} = \frac{d}{dt} \int_0^L \eta \, dx \geq [q(L, t) - q(0, t)]/\theta_B, \tag{2.1.12}$$

agreeing with the Clausius–Planck inequality (1.2.2). Or, we might replace these temperature boundary conditions by boundary conditions of the radiation type. With Newton's law of cooling, this would give

$$\begin{aligned} q(L, t) &= \alpha[\theta_B - \theta(L, t)], \\ q(0, t) &= -\alpha[\theta_B - \theta(0, t)] \end{aligned} \tag{2.1.13}$$

with α a positive constant. That α should be positive reflects the idea that heat should flow into the bar if it is cooler than the heat bath, out if it is hotter. Then, it is easy to verify that

$$\begin{aligned} q(L, t)/\theta(L, t) &= \alpha[\theta_B - \theta(L, t)]/\theta \geq q(L, t)/\theta_B \\ -q(0, t)/\theta(0, t) &\geq -q(0, t)/\theta_B. \end{aligned} \tag{2.1.14}$$

Putting this together with (2.1.4) and (2.1.9), again assuming $r = 0$, we get

$$\frac{dS}{dt} \geq Q/\theta_B, \tag{2.1.15}$$

also agreeing with (1.2.2). Another commonly used radiation condition is the Stefan–Boltzmann law, which would give

$$\begin{aligned} q(L, t) &= \beta[\theta_B^4 - \theta^4(L, t)], \\ q(0, t) &= -\beta[\theta_B^4 - \theta^4(0, t)] \end{aligned} \tag{2.1.16}$$

with β a positive constant, again leading to (2.1.15) when $r = 0$.

One can consider heat transfer from the sides of the bar while insulating the ends. Using, for example, Newton's law of cooling, this gives

$$q(0, t) = q(L, t) = 0, \qquad r = \alpha[\theta_B - \theta(x, t)]. \tag{2.1.17}$$

Again, α should be positive, to have heat flow into the bar when it is cooler and out of the bar when it is hotter. Also, it follows that

$$r(x, t)/\theta(x, t) \geq r(x, t)/\theta_B(t), \tag{2.1.18}$$

and that (2.1.15) again holds, now with $Q = Q_b$. One can replace the insulated end conditions by radiation boundary conditions, again confirming (2.1.15). So, under various assumptions of this kind, the Clausius–Duhem inequality implies the Clausius–Planck inequality. Unlike the latter, it also applies to cases where θ_B might be considered to depend on x, or to cases where θ might be assigned different values at the ends, in a bar with insulated sides.

For classical studies of equilibrium, one needs a constitutive equation depending on the material, and one of the form

$$\varepsilon = \varepsilon(\eta) \qquad (2.1.19)$$

would fit Gibbs's preference. Let us consider this bar in contact with a heat bath at temperature $\theta_B = $ const. and explore the consequences of definition III, in Section 1.3. We have

$$E_B = \int_0^L [\varepsilon(\eta) - \theta_B \eta]\, dx, \qquad (2.1.20)$$

defined for some functions $\eta(x)$. At least one of these should occur as a thermodynamic equilibrium state; call it $\bar{\eta}$. As possible variations of it, we consider a one-parameter family of functions, as is commonly done in the calculus of variations,[2]

$$\eta = \bar{\eta} + \mu \delta\eta,$$

where μ is the parameter and $\delta\eta$ an arbitrary smooth function of x. A common idea is that any variation of this kind is possible, if it is small enough, and, given $\delta\eta$, we can make it small by making μ small. For the moment, consider $\delta\eta$ as fixed, and consider

$$E_B(\mu) = \int_0^L [\varepsilon(\bar{\eta} + \mu \delta\eta) - \theta_B(\bar{\eta} + \mu \delta\eta)]\, dx. \qquad (2.1.21)$$

For μ small, we have, as a first approximation,

$$E_B(\mu) \cong E_B(0) + \mu \delta E_B,$$

where

$$\delta E_B = E'_B(0) = \int_0^L \left[\frac{d\varepsilon}{d\eta}(\bar{\eta}) - \theta_B\right] \delta\eta\, dx \qquad (2.1.22)$$

is called the *first variation*. We interpret the criterion for equilibrium as meaning that we should have

$$\mu \delta E_B \geq 0,$$

---

[2]One of the better elementary treatments of this subject is reference [11].

and, since μ can be chosen to be positive or negative, this requires that

$$\delta E_B = 0,$$

for arbitrary $\delta\eta$. By a fundamental lemma in the calculus of variations, this holds if and only if

$$\frac{d\varepsilon}{d\eta}(\bar\eta) = \theta_B. \tag{2.1.23}$$

Briefly, if this were not true, the bracketed quantity in (2.1.22) would be positive or negative in some interval. Choosing $\delta\eta$ positive in a subinterval of this and zero elsewhere produces a contradiction. So, (2.1.23) gives an equation for determing (stable or unstable) equilibrium values of η. With other theories, similar reasoning is used in interpreting the criteria for equilibrium. In equilibrium, one expects θ, the temperature of the bar, to match that of the heat bath. This fits the common assumption that θ and η are always related by

$$\theta = \frac{d\varepsilon}{d\eta}. \tag{2.1.24}$$

As a general proposition, it is certainly not safe to assume that an equation that holds in equilibrium also holds more generally. Here, there is a consensus of opinion that the assumption has a range of validity which is not restricted to equilibria but might well not include processes involving large departures from equilibrium.

Now, we can proceed to obtain stability conditions. As a test for a local minimum, many would use the second derivative test,

$$\delta^2 E_B = E_B''(0) \geq 0 \tag{2.1.25}$$

or

$$\int_0^L \frac{d^2\varepsilon}{d\eta^2}(\bar\eta)(\delta\eta)^2\,dx \geq 0$$

which, by (2.1.24), leads to

$$\frac{d\theta}{d\eta}(\bar\eta) = \frac{d^2\varepsilon}{d\eta^2}(\bar\eta) \geq 0. \tag{2.1.26}$$

Briefly, if the second derivative were negative, it would be negative in some interval by continuity and one could again find $\delta\eta$ for which the integral inequality is violated.

Conditions very similar to (2.1.23) and (2.1.26) emerge, by similar reasoning, in a great variety of theories. They represent what may be viewed as minimal requirements for stable or metastable equilibrium. If (2.1.24) gives θ as a monotonically increasing function of η, taking on all positive values, then (2.1.23) holds for exactly one value of $\bar\eta$ and (2.1.26) will be satisfied. Some workers are happy to consider only constitutive equations having these properties. This excludes some instabilities that might occur in principle, but I know of no clear evidence that they occur in practice.

According to the criterion, the most stable equilibria correspond to absolute minima. Clearly, the integral will be smallest if the integrand everywhere takes on its minimum value,[3] so the criterion for this is

$$\varepsilon(\eta) - \theta_B \eta \geq \varepsilon(\bar{\eta}) - \theta_B \bar{\eta},$$

or with (2.1.23),

$$\varepsilon(\eta) - \varepsilon(\bar{\eta}) \geq \frac{d\varepsilon}{d\eta}(\bar{\eta})(\eta - \bar{\eta}) \supset \eta, \qquad (2.1.27)$$

where $\supset$ is to be read as "for all." The aforementioned common assumption implies that this holds for every choice of η and $\bar{\eta}$. Mathematically, ε is then a convex function of η. Were it not, (2.1.27) would require that $(\bar{\eta}, \varepsilon(\bar{\eta}))$ be a point of convexity of the graph of $\varepsilon = \varepsilon(\eta)$; the graph must be above the tangent line at this point.

For simplicity, make the assumption alluded to above, so we can solve (2.1.24) to obtain η as a function of θ, either quantity serving as a label for states. The derivatives will satisfy

$$\frac{d\eta}{d\theta} \frac{d\theta}{d\eta} = 1.$$

Note that, if equality holds in (2.1.26), $d\eta/d\theta$ will not be finite. Actually, this can happen at particular values of θ at which phase transitions take place, although some more general theory is needed to describe what then occurs.[4] We will exclude this possibility.

Now consider starting with a bar in equilibrium at some temperature θ_B. Change this a little, to $\theta_B + \Delta\theta_B$, and allow it to reach equilibrium. This will give a change in energy; by (2.1.23) and (2.1.24)

$$\Delta E = \Delta \int_0^L \varepsilon(\bar{\eta}) \, dx$$

$$\cong L \frac{\overline{d\varepsilon}}{d\theta} \Delta\theta_B \cong L\theta_B \frac{\overline{d\eta}}{d\theta} \Delta\theta_B. \qquad (2.1.28)$$

Here, the overbar denotes evaluation at the state corresponding to $\bar{\eta}$. Since no work is done on the bar, this should also be the total amount of heat supplied to the bar, which can be measured using a calorimeter. Clearly L, θ_B and $\Delta\theta_B$ can also be measured, so one can infer values of $d\varepsilon/d\theta$ and $d\eta/d\theta$. By doing this for a range of temperatures, one can get ε and

---

[3]This excludes some commonly ignored mathematical possibilities, like the function not being bounded below, or not attaining a minimum.

[4]See, for example, the discussion of lambda transitions by Pippard [4].

η as functions of θ, at least for some range of temperatures, and from this, calculate $\varepsilon(\eta)$. A careful worker will at least increase and decrease temperature to see if the results thus obtained are consistent.

2.2   Constitutive Theory for Rigid Bars

In the previous discussion, we said rather little about some items, q in particular, but they play a more important role in the consideration of heat transfer. The aim is to somehow relate these to temperature. The classical theory of heat conduction uses Fourier's law

$$q = \kappa \, \partial\theta/\partial x, \tag{2.2.1}$$

where κ is a constant, or more realistically, a function of temperature. This suggests that items of interest should depend on θ and $\partial\theta/\partial x$. Equilibrium studies suggest that ε and η depend only on θ, but they deal with cases where $\partial\theta/\partial x$ could well be small enough to make a negligible contribution. Thus, a likely assumption may be that we have constitutive equations of the form

$$\varepsilon = \varepsilon(\theta, \partial\theta/\partial x), \quad \eta = \eta(\theta, \partial\theta/\partial x), \quad q = q(\theta, \partial\theta/\partial x), \tag{2.2.2}$$

depending on the material from which the bar is made. The usual idea is that these do not change when we put the bar in different environments.

In a slightly different category is r, which is not as independent of the environment from the examples of radiation laws. We do need some prescription(s) for it, more or less like the examples discussed in Section 2.1. The basic idea is to convert the energy equation (2.1.8) to a differential equation for θ and to arrange that any solutions satisfy the Clausius–Duhem inequality (2.1.10). One way of proceeding is to eliminate r between the two to obtain

$$\theta \frac{\partial\eta}{\partial t} - \theta \frac{\partial}{\partial x}(q, \theta) \geq \frac{\partial\varepsilon}{\partial t} - \frac{\partial q}{\partial x} \tag{2.2.3}$$

and to restrict the constitutive equations (2.2.2), so this holds for essentially arbitrary smooth functions $\theta(x, t)$; one can respect obvious conditions such as $\theta > 0$. This procedure is not as arbitrary as it may seem. For example, suppose that we accept Newton's law of cooling together with the prescription for r given in (2.1.17) with the idea that θ_B is an assignable constant. Choose the bar temperature function $\theta(x, t)$ and think of using (2.2.2) to calculate $\varepsilon(x, t)$, and so on. Then calculate

$$f(x, t) = \frac{\partial\varepsilon}{\partial t} - \frac{\partial q}{\partial x}.$$

Now choose particular values of $(x, t) = (x_0, t_0)$, say, and solve

$$f(x_0, t_0) = r(x_0, t_0) = \alpha[\theta_B - \theta(x_0, t_0)]$$

for θ_B, to satisfy (2.1.8) at (x_0, t_0). Clearly, (2.2.3) must then hold at (x_0, t_0). Actually, it is not too unrealistic to assume that θ_B is an assignable function of x, t. Accept this and avoid evaluation at particular values of x and t. A weakness in the argument is that the calculated values of θ_B could be negative, and this is not physically acceptable. Hence, we may sacrifice some generality in accepting the indicated assumption.

We now rearrange (2.2.3) by introducing

$$\phi = \varepsilon - \theta\eta = \phi\left(\theta, \frac{\partial\theta}{\partial x}\right),\qquad(2.2.4)$$

the Helmholtz free energy per unit length. By simple calculation,

$$\frac{\partial\phi}{\partial t} = \frac{\partial\phi}{\partial\theta}\frac{\partial\theta}{\partial t} + \frac{\partial\phi}{\partial(\partial\theta/\partial x)}\frac{\partial^2\theta}{\partial x\partial t}$$

$$\leq q\frac{\partial\theta}{\partial x}\Big/\theta - \eta\frac{\partial\theta}{\partial t}.\qquad(2.2.5)$$

We want to ensure this will hold for any choice of the function $\theta(x, t)$, assuming (2.2.2) is used to calculate q. Consider $(x, t) = (x_0, t_0)$, any particular point and time. It is easy to construct temperature functions such that

$$\theta(x_0, t_0) = a,$$

$$\frac{\partial\theta}{\partial x}(x_0, t_0) = b,$$

$$\frac{\partial\theta}{\partial t}(x_0, t_0) = c,\qquad(2.2.6)$$

$$\frac{\partial^2\theta}{\partial x\partial t}(x_0, t_0) = d,$$

where a, b, c and d are arbitrary constants, with a > 0; simple polynomial functions will suffice.

Now, use these values in (2.2.5) evaluated at (x_0, t_0), to obtain an inequality of the form

$$A + Bc + Cd \leq 0,$$

where A, B and C are functions of a and b only. Fix a, b, and c. If $C > 0$, we could take d negative and large enough to violate this inequality and, if $C < 0$, we could similarly violate it. A similar argument applies to B, so we must have

$$B = C = 0,\qquad A \leq 0,$$

for all values of a and b, that is, for all values of θ and $\partial\theta/\partial x$. This gives

$$\frac{\partial\phi}{\partial\left(\frac{\partial\theta}{\partial x}\right)} \equiv 0 \Rightarrow \phi = \phi(\theta),\qquad(2.2.7)$$

and

$$\eta = -\frac{d\phi}{d\theta}.$$ (2.2.8)

This leaves us with the inequality

$$g\left(\theta, \frac{\partial\theta}{\partial x}\right) = q\frac{\partial\theta}{\partial x} \geq 0.$$ (2.2.9)

Now, for θ fixed, g clearly vanishes when $\partial\theta/\partial x = 0$, and, being nonnegative, it has a minimum there, so

$$\frac{\partial g}{\partial\left(\frac{\partial\theta}{\partial x}\right)}(\theta, 0) = q(\theta, 0) = 0.$$ (2.2.10)

Physically, θ is constant in equilibrium and the heat flux then vanishes. Fourier's law (2.2.1) then emerges in a natural way as a first approximation for $\partial\theta/\partial x$ suitably small. Equation (2.2.9) reduces to the condition $\kappa \geq 0$, which is always assumed in such theory. With a more nonlinear theory, q must be chosen to satisfy (2.2.9).

As was mentioned in our considerations of equilibrium theory, it is commonly assumed that η is a monotonically increasing function of θ, so, by (2.2.8)

$$\frac{d\eta}{d\theta} = -\frac{d^2\phi}{d\theta^2} > 0,$$ (2.2.11)

and (2.2.8) can be inverted, to obtain θ as a function of η.

Also, from (2.2.4) and (2.2.8),

$$\frac{d\varepsilon}{d\theta} = \frac{d}{d\theta}(\phi + \theta\eta) = \frac{d\phi}{d\theta} + \eta + \theta\frac{d\eta}{d\theta}$$

$$= \theta\frac{d\eta}{d\theta},$$

involved earlier in (2.1.28), there deduced from equilibrium theory. With this, the energy equation (2.1.8) becomes

$$\frac{\partial\varepsilon}{\partial t} = \frac{d\varepsilon}{d\theta}\frac{\partial\theta}{\partial t} = \theta\frac{d\eta}{d\theta}\frac{\partial\theta}{\partial t}$$

$$= \theta\frac{\partial\eta}{\partial t} = \frac{\partial q}{\partial x} + r.$$ (2.2.12)

With, say, Fourier's law and Newton's law of cooling, this gives

$$\frac{d\varepsilon}{d\theta}\frac{\partial\theta}{\partial t} = \frac{\partial}{\partial x}\left(\kappa(\theta)\frac{\partial\theta}{\partial x}\right) + \alpha(\theta_B - \theta),$$ (2.2.13)

as the temperature equation, our discussion in Section 2.1 covering some possible boundary conditions. Linearizing this about $\theta = \theta_B = \text{const.}$, gives a linear equation, likely to be used in problem solving in cases where the change in temperature is expected to be small compared to some constant value. The linearized equation is of the form

$$C_0 \frac{\partial \theta}{\partial t} = \kappa_0 \frac{\partial^2 \theta}{\partial x^2} + \alpha(\theta_B - \theta), \tag{2.2.14}$$

with

$$C_0 = \frac{d\varepsilon}{d\theta}(\theta_B), \qquad \kappa_0 = \kappa(\theta_B). \tag{2.2.15}$$

Equation (2.2.14), with $\alpha = 0$, is known as the *diffusion equation* or the *heat equation*.

If (2.2.12) holds, we can calculate that

$$\frac{dS}{dt} = \frac{d}{dt} \int_0^L \eta \, dx = \frac{q}{\theta}\Big|_0^L + \int_0^L \left(\frac{r}{\theta} + q \frac{\partial \theta}{\partial x} \Big/ \theta^2 \right) dx. \tag{2.2.16}$$

With this and the earlier discussion of heat baths, it is fairly easy to convince yourself that the assumptions of reversibility made in calorimetry cannot be quite correct according to this kind of nonequilibrium theory. Generally, the nonequilibrium theory is considered to be sounder. This and similar experiences with other nonequilibrium theories serve to motivate the rather common opinion that reversible processes are not really physically attainable. There is one case in which a simple analysis gives an estimate of the errors involved. Suppose we start with the bar in equilibrium, with

$$\theta = \theta_B^0 = \theta_0 \quad \text{for } t < 0. \tag{2.2.17}$$

Suppose that we insulate the ends, so that

$$q(L, t) = q(0, t) = 0, \supset t. \tag{2.2.18}$$

Then, at $t = 0$, we change the heat bath temperature to

$$\theta_B = (1 + \sigma)\theta_0, \tag{2.2.19}$$

where σ is a small number. Then, reasonably, we can use the linearized equation (2.2.14), with boundary conditions

$$\kappa_0 \frac{\partial \theta}{\partial x} = 0 \quad \text{at } x = 0, L,$$

and we can satisfy this by taking θ independent of x. This would also be true if we used nonlinear theory. However, with linear theory, we get a simple equation,

$$\frac{\partial \theta}{\partial t} = \beta(\theta_B - \theta), \qquad \beta = \frac{\alpha}{C_0} > 0.$$

Integrating this, and using the initial condition (2.2.17), we get

$$\theta_B - \theta = (\theta_B - \theta_0)\exp(-\beta t)$$
$$= \sigma\theta_0 \exp(-\beta t), \tag{2.2.20}$$

so θ approaches θ_B exponentially, as $t \to \infty$. The total heat supplied to the bar is

$$\delta Q = \int_0^L \int_0^\infty r\, dt\, dx = \alpha L \int_0^\infty (\theta_B - \theta)\, dt \tag{2.2.21}$$

$$= L\,\frac{\alpha\sigma\theta_0}{\beta} = LC_0\sigma\theta_0.$$

With (2.2.16) the total change in entropy, according to nonlinear theory, is

$$\Delta S = \int_0^L \int_0^\infty \frac{r}{\theta}\, dt\, dx = \alpha L \int_0^\infty \frac{\theta_B - \theta}{\theta}\, dt. \tag{2.2.22}$$

As estimated by linear theory, we calculate that

$$\frac{\theta_B - \theta}{\theta} = \frac{\sigma \exp(-\beta t)}{1 + \sigma[1 - \exp(-\beta t)]}$$
$$\cong \sigma \exp(-\beta t)$$

with an error of the order $O(\sigma^2)$. In this approximation, we then obtain

$$\Delta S \cong \frac{L\alpha\sigma}{\beta} = \frac{\delta Q}{\theta_0},$$

or the approximate reversibility needed to correlate with equilibrium studies such as are involved in calorimetry. So, one needs βt to be large to get near thermodynamic equilibrium and the heat bath temperature changes to be small, as measured by σ, to approximate results obtained from equilibrium theory.

This exemplifies a remark made in Chapter 1. The desired reversible processes do not always exist. Often, we can approximate reversibility as closely as we like, here by making σ small enough. In the limit $\sigma = 0$, we do get a reversible process, but it requires the bar to remain in equilibrium at θ θ_B, and, for calorimetry, we need the temperature to change slightly.

2.3   Constitutive Theory for Thermoelastic Bars

Here we generalize the previous theory, allowing the bar to stretch. This involves picking a reference configuration of the bar. The usual practice is to identify this as a rather stable equilibrium configuration, subject to no forces, at a particular (constant) reference temperature θ_R. Here, x will describe positions in this reference configuration, a fixed interval

$$0 \le x \le L, \tag{2.3.1}$$

as before. Such points can move to other points of the line containing this interval. Motion will then be described by an equation of the form

$$y = y(x, t), \tag{2.3.2}$$

where y denotes the coordinate of a material "point," really an average position of a cross section which was at x in the reference configuration.

Two quantities of interest are the velocity

$$\dot{y} = \frac{\partial y}{\partial t} \tag{2.3.3}$$

and the stretch[5]

$$\lambda = \frac{\partial y}{\partial x} > 0, \tag{2.3.4}$$

a dimensionless measure of changes of length. Generally, we will use a superposed dot to denote partial differentiation with respect to t, holding x fixed. Often, workers use instead the strain,

$$e = \lambda - 1, \tag{2.3.5}$$

interpretable as the change of length divided by the reference length of an infinitesimal part of the bar. As before, we assume that the physical properties of one cross section are like those of any other. Physically, the bar should be made of an homogeneous material and be of cylindrical form. Then A_R, the area of a cross section in the reference configuration, is a constant, unaffected by motions of the bar. If one is dealing with bars of the same material and geometrically similar in shape, it is commonly assumed that they can be described by essentially the same equations and this is in reasonably good agreement with experience. Certainly, a fatter bar has a greater mass, for example. To compensate for such obvious differences, we consider additive quantities, such as energy, entropy and momentum, as divided by A_R. Thus, in applying the first law to parts of a bar, we use

$$\overline{E} = \overline{P} + \overline{Q}, \quad \overline{E} = E/A_R, \quad \overline{P} = P/A_R, \quad \overline{Q} = Q/A_R. \tag{2.3.6}$$

Since it becomes clumsy to introduce too many notations we will drop the bars, understanding that E is to be thought of as energy per unit reference area, and so on.

---

[5]The value $\lambda > 0$ reflects the physical idea that the length of a segment cannot be reduced to zero.

Equations of motion are based on the idea that the rate of change of momentum equals the force. With one-dimensional theory, we can only account for forces directed along the line representing the bar. Detailing this, we introduce the format

$$\frac{d}{dt} \int_{x_1}^{x_2} \rho \dot{y} \, dx = \int_{x_1}^{x_2} f \, dx + \sigma \Big|_{x_1}^{x_2}, \tag{2.3.7}$$

with x_1 and x_2 representing any two points in the interval (2.1.1). Here ρ is the (constant) mass per unit reference volume, so

$$\int_{x_1}^{x_2} \rho \dot{y} \, dx$$

represents momentum per unit reference area. Similarly, f represents force per unit reference volume, distributed along the length of the bar. With σ we denote what is commonly called the engineering, or Piola–Kirchhoff stress. It covers the force exerted on the material on one side of a cross section by the nearby material on the other side. Adopting arguments very similar to those used to deduce (2.1.8), we obtain from (2.3.7) the partial differential equation

$$\rho \ddot{y} = \rho \frac{\partial^2 y}{\partial t^2} = f + \frac{\partial \sigma}{\partial x}. \tag{2.3.8}$$

With the motion, the energy will consist in part of kinetic energy and it is convenient to recognize this explicitly. So we write, for the energy (per unit reference area), for a part of the bar

$$E = \int_{x_1}^{x_2} \left(\varepsilon + \frac{1}{2} \rho \dot{y}^2 \right) dx. \tag{2.3.9}$$

Here ε is called the *internal energy* (per unit reference volume). Similarly, we need to consider the power represented by

$$P = \int_{x_1}^{x_2} f \dot{y} \, dx + \sigma \dot{y} \Big|_{x_1}^{x_2} = \int_{x_1}^{x_2} \left[f \dot{y} + \frac{\partial}{\partial x} (\sigma \dot{y}) \right] dx. \tag{2.3.10}$$

For Q, we use the same kind of format adopted for rigid bars,

$$Q = \int_{x_1}^{x_2} r \, dx + q \Big|_{x_1}^{x_2} = \int_{x_1}^{x_2} \left(r + \frac{\partial q}{\partial x} \right) dx. \tag{2.3.11}$$

Bear in mind that this represents the rate at which heat is supplied, per unit reference area, so this q has different physical dimensions. Then, by arguments which should now seem familiar, we use the first law,

$$\frac{dE}{dt} = P + Q,$$

to obtain the differential equation

$$\overline{\dot{\varepsilon} + \tfrac{1}{2}\rho\dot{y}^2} = \dot{\varepsilon} + \rho\dot{y}\,\ddot{y}$$

$$= f\dot{y} + \frac{\partial}{\partial x}(\sigma\dot{y}) + r + \frac{\partial q}{\partial x}.$$

Using (2.3.8), we can cancel some terms giving

$$\dot{\varepsilon} = \sigma\frac{\partial\dot{y}}{\partial y} + r + \frac{\partial q}{\partial x}, \tag{2.3.12}$$

the idea being to use this as an equation for $\theta(x, t)$.

Again, we use the Clausius–Duhem inequality (2.1.9),

$$\frac{dS}{dt} = \frac{d}{dt}\int_{x_1}^{x_2} \eta\, dx \geq \int_{x_1}^{x_2} \frac{r}{\theta}\, dx + \frac{q}{\theta}\bigg|_{x_1}^{x_2}, \tag{2.3.13}$$

merely reinterpreting S to be entropy per unit reference area. As before, we then have

$$\dot{\eta} \geq \frac{1}{\theta}\left(r + \frac{\partial q}{\partial x}\right) - q\frac{\partial\theta}{\partial x}\bigg/\theta^2. \tag{2.3.14}$$

The usual idea is that, for a given bar, f and r may be specified in different ways, depending on the systems with which it is to be in contact. So, we require a theory enabling rather arbitrary specifications of these and we will not formulate constitutive equations for them. In (2.3.12), f has already been eliminated and, as before, we can eliminate r between (2.3.12) and (2.3.14), to obtain the inequality

$$\dot{\varepsilon} - \theta\dot{\eta} \leq \sigma\frac{\partial\dot{y}}{\partial x} + (q/\theta)\frac{\partial\theta}{\partial x}.$$

Also, as before, it is convenient to introduce

$$\phi = \varepsilon - \theta\eta, \tag{2.3.15}$$

the Helmholtz free energy per unit reference volume, satisfying

$$\dot{\phi} \leq \sigma\frac{\partial\dot{y}}{\partial x} - \eta\dot{\theta} + (q/\theta)\frac{\partial\theta}{\partial x}. \tag{2.3.16}$$

Then, we seek constitutive relations, relating ϕ, σ, η, and q to motion and temperature, allowing for the fact that different bars can respond differently. Here, we aim at theories which are, in a sense, very local. Consider any fixed values of x, t, designated by x_0 and t_0. For small enough values of $x - x_0$ and $t - t_0$, we have, approximately

$$y(x, t) \cong y(x_0, t_0) + \frac{\partial y}{\partial x}(x_0, t_0)(x - x_0) + \dot{y}(x_0, t_0)(t - t_0),$$

$$\tag{2.3.17}$$

$$\theta(x, t) \cong \theta(x_0, t_0) + \frac{\partial\theta}{\partial x}(x_0, t_0)(x - x_0) + \dot{\theta}(x_0, t_0)(t - t_0).$$

Then, roughly, the assumption is that the behavior of a material point is influenced by the general motion and temperature. This influence is small except when it occurs very close to the point in position and time. Using (2.3.17) to estimate this, we are led to the assumption that ϕ, σ, η, and q are functions of the following quantities:

$$y, \quad \lambda = \frac{\partial y}{\partial x}, \quad \dot{y}, \quad \theta, \quad \frac{\partial \theta}{\partial x}, \quad \dot{\theta}. \tag{2.3.18}$$

Two arguments are used to restrict these functions. One is the notion that these should be objective functions, as it is sometimes put, meaning that their values are unaffected by superposing rigid motions. Suppose, for example, that we have two motions $y(x,t)$ and $\bar{y}(x,t)$, with

$$\bar{y}(x,t) = y(x,t) + a + bt,$$

with a and b constant. Here, it would be acceptable to replace $a + bt$ by an arbitrary smooth function of t, but one gets the same restrictions, either way. Suppose also that the temperature functions are the same,

$$\bar{\theta}(x,t) = \theta(x,t).$$

It is true that a given particle takes on different positions and the velocity differs by a constant; but, physically, we do not expect this to affect heat fluxes, entropies, and so on. Thus, for example, we should have

$$\phi \left(y + a + bt, \ \frac{\partial y}{\partial x}, \ \dot{y} + b, \ \theta, \ \frac{\partial \theta}{\partial x}, \ \dot{\theta} \right)$$
$$= \phi \left(y, \ \frac{\partial y}{\partial x}, \ \dot{y}, \ \theta, \ \frac{\partial \theta}{\partial x}, \ \dot{\theta} \right),$$

for any given motion $y(x,t)$ and any choice of constants a and b. It then follows that this will be true only if ϕ does not depend on y or $\dot{y}$. By this kind of argument, we reduce (2.3.18) to

$$\frac{\partial y}{\partial x}, \quad \theta, \quad \frac{\partial \theta}{\partial x}, \quad \dot{\theta}. \tag{2.3.19}$$

Assuming ϕ, η, σ, and q depend on these variables, we put this into (2.3.16) to get

$$\left[\frac{\partial \phi}{\partial (\partial y/\partial x)} - \sigma \right] \frac{\partial \dot{y}}{\partial x} + \left(\frac{\partial \phi}{\partial \theta} + \eta \right) \dot{\theta} + \frac{\partial \phi}{\partial \dot{\theta}} \ddot{\theta}$$
$$+ \frac{\partial \phi}{\partial (\partial \theta/\partial x)} \frac{\partial \dot{\theta}}{\partial x} \leq (q/\theta) \frac{\partial \theta}{\partial x}. \tag{2.3.20}$$

As before, we want to restrict the constitutive functions so that the inequality is satisfied for all motions and temperature distributions. We use the fact that, with the variables (2.3.19) fixed, at $x = x_0$, $t = t_0$, we can still regard as independent variables higher derivatives such as $\ddot{\theta}$. This gives the restrictions

$$\phi = \phi\left(\frac{\partial y}{\partial x}, \theta\right) = \phi(\lambda, \theta)$$

$$\sigma = \frac{\partial \phi}{\partial(\partial y/\partial x)} = \frac{\partial \phi}{\partial \lambda}$$

(2.3.21)

with the remaining restriction that

$$(q/\theta)\frac{\partial \theta}{\partial x} - \left(\frac{\partial \phi}{\partial \theta} + \eta\right)\dot{\theta} \geq 0.$$

(2.3.22)

Since this expression vanishes when $\dot{\theta} = \partial\theta/\partial x = 0$, it then takes on its minimum value. Applying the first derivative test for a minimum to this function then gives

$$\dot{\theta} = \partial\theta/\partial x = 0 \Rightarrow q = 0, \qquad \eta = -\partial\phi/\partial\theta.$$

(2.3.23)

Usually, workers assume that q is independent of $\dot{\theta}$ and that η is independent of $\dot{\theta}$ and $\partial\theta/\partial x$, and we will follow suit. With the indicated assumption we have, always,

$$\eta = -\partial\phi/\partial\theta, \qquad q\left(\frac{\partial y}{\partial x}, \theta, \frac{\partial \theta}{\partial x}\right)\frac{\partial \theta}{\partial x} \geq 0.$$

(2.3.24)

Often this is specialized more, with the assumption that

$$q = k\left(\frac{\partial y}{\partial x}, \theta\right)\frac{\partial \theta}{\partial x}, \qquad k \geq 0,$$

(2.3.25)

which seems to be adequate practically. Underlying (2.3.24) is the classical view that entropy is defined using processes which are, at least approximately, reversible. It would not hurt for you to think hard about this yourself, but it does at least seem difficult to reconcile this with the idea that η can depend on $\dot{\theta}$ and/or $\partial\theta/\partial x$.[6] The effect of this is to assume that, at each particle, the entropy is given by the constitutive equation applying to equilibria.

For various other theories which are, in a similar sense, local, thermodynamicists often use this assumption with some success. It is called the *principle of local states* or the *hypothesis of local thermodynamic equilibrium* and other similar names. Often workers interested in viscoelastic effects employ theories which are not local, in the temporal sense. For example, the

---

[6]If you wish to explore this more deeply, you might find it helpful first to study Chapter 10.

stress at any particular time depends on values of the stretch experienced at all previous times. Here, such assumptions are commonly rejected. For example, one expert, Rivlin [12] offers physical arguments to help explain why he considers them to be unsound for use in connection with such theories. Various other experts have made clear that they came to a similar conclusion. This will be discussed in more detail in Chapter 10. It is not reasonable to infer from this that the reasons that convinced one would be the same as those that convince another. Obviously, one needs some alternative, and, concerning this, there are some differences of opinion. There are still different ideas about entropy, stemming from statistical molecular theories, for example, which do not require the association with reversible processes. Such theory also suggests a need for generalizing the Clausius–Duhem inequality. It would take a very lengthy discussion to elaborate these remarks so we will not pursue it. With the specialization indicated, we have what is generally regarded as thermoelasticity theory or, more properly, the one-dimensional version of such theory. Such theory has been used, successfully, to describe numerous phenomena.

With σ given by (2.3.21), the equations of motion (2.3.8) become

$$\rho\ddot{y} = \frac{\partial}{\partial x}\left(\frac{\partial\phi}{\partial\lambda}\right) + f, \qquad (2.3.26)$$

it being understood that some appropriate prescription for f is to be given. In most situations, it is reasonable to take $f = 0$. When exceptions are encountered, they are likely to involve forces exerted by gravitational or electromagnetic fields and, in the latter case, it is probable that more general kinds of constitutive equations are needed.

To obtain the temperature equation, we use (2.3.12) and note that with our constitutive equations,

$$\dot{\varepsilon} = \overline{\dot{\phi} + \theta\eta} = \left(\frac{\partial\phi}{\partial\theta} + \eta\right)\dot{\theta} + \theta\dot{\eta} + \frac{\partial\phi}{\partial\lambda}\dot{\lambda}$$

$$= \sigma\dot{\lambda} + \frac{\partial q}{\partial x} + r$$

$$= \frac{\partial\phi}{\partial\lambda}\dot{\lambda} + \frac{\partial q}{\partial x} + r$$

from which

$$\theta\dot{\eta} = -\theta\left(\overline{\frac{\partial\phi}{\partial\theta}}\right) = \frac{\partial q}{\partial x} + r. \qquad (2.3.27)$$

Superficially, this is the same as the equation (2.2.12) derived for rigid bars. However, here η depends not only on θ, but also on λ, so (2.3.26) and (2.3.27) are a coupled set of equations for motion and temperature. Again, r needs to be specified and the possibilities for this are well-covered by our discussion of rigid bars.

Sometimes (2.3.21) is replaced by an alternative. Suppose we can invert (2.3.23) to get

$$\theta = f(\eta, \lambda).$$

Then, express ε as a function of λ and η, as indicated by

$$\varepsilon = \phi(\lambda, \theta) + \theta\eta(\lambda, \theta)$$
$$= \phi(\lambda, f(\eta, \lambda)) + f(\eta, \lambda)\eta = \varepsilon(\lambda, \eta).$$

Mathematically, ε is a Legendre transform of ϕ and such transforms are used in various places in thermodynamics. Then, by the chain rule,

$$\left.\frac{\partial\varepsilon}{\partial\eta}\right|_\lambda = \frac{\partial\phi}{\partial\theta}\frac{\partial f}{\partial\eta} + \theta + \eta\frac{\partial f}{\partial\eta} = \theta$$

and

$$\left.\frac{\partial\varepsilon}{\partial\lambda}\right|_\eta = \left.\frac{\partial\phi}{\partial\lambda}\right|_\theta + \frac{\partial\phi}{\partial\theta}\frac{\partial f}{\partial\lambda} + \eta\frac{\partial f}{\partial\lambda}$$
$$= \left.\frac{\partial\phi}{\partial\lambda}\right|_\theta = \sigma.$$

So, instead of (2.3.21), we can use

$$\varepsilon = \varepsilon(\lambda, \eta),$$

$$\sigma = \frac{\partial\varepsilon}{\partial\lambda}, \tag{2.3.28}$$

$$\theta = \frac{\partial\varepsilon}{\partial\eta}.$$

Various kinds of wave propagation are considered to take place almost isentropically (η nearly constant), and, for such processes, (2.3.28) is a convenient choice. Processes more nearly isothermal are encountered in slow, quasi-static tests done at constant temperature and, for these, (2.3.21) is more convenient.

It is worth noting that, while $\phi(\lambda, \theta)$ and $\varepsilon(\lambda, \eta)$ are, in the sense indicated, equivalent constitutive equations, an equation of the form

$$\varepsilon = \varepsilon(\lambda, \theta)$$

gives a less complete description. Suppose, for example, that this function is independent of λ:

$$\varepsilon = \varepsilon(\theta). \tag{2.3.29}$$

This statement does in fact apply rather well to various kinds of rubber as long as λ is not extremely large. When (2.3.29) holds, we have, using

(2.3.24),

$$\frac{d\varepsilon(\theta)}{d\theta} = \frac{\partial}{\partial\theta}(\phi + \theta\eta)$$

$$= \frac{\partial\phi}{\partial\theta} + \eta + \theta\frac{\partial\eta}{\partial\theta} = \theta\frac{\partial\eta}{\partial\theta}.$$

This implies that η is of the form

$$\eta = f(\theta) + g(\lambda), f' = \varepsilon'(\theta)/\theta. \tag{2.3.30}$$

Also,

$$\frac{\partial\varepsilon}{\partial\lambda} = 0 = \frac{\partial}{\partial\lambda}(\phi + \theta\eta)$$

$$= \sigma + \theta\frac{\partial\eta}{\partial\lambda}$$

$$= \sigma + \theta g'(\lambda).$$

or

$$\sigma = -\theta\frac{\partial\eta}{\partial\lambda} = -\theta g'(\lambda). \tag{2.3.31}$$

Also,

$$\phi = h(\theta) - \theta g(\lambda), \tag{2.3.32}$$

where

$$h(\theta) = \varepsilon(\theta) - \theta f(\theta).$$

Given any function ϕ of the form (2.3.32), we have

$$\varepsilon = \phi + \theta\eta = \phi - \theta\frac{\partial\phi}{\partial\theta} = h - \theta h',$$

which is a function only of θ. Clearly, we can also obtain σ and η from (2.3.32). However, from (2.3.29), there is no way to determine $g(\lambda)$. Response of this kind is sometimes called "entropy elasticity" since, from (2.3.31), σ is associated primarily with changes in η and this indicates a characteristic feature: at fixed λ, σ is proportional to θ.

2.4   Constitutive Theory for Shearing of Plates

By what is essentially a reinterpretation of the foregoing theory, we can obtain the one-dimensional theory of shearing of plates. Here, we think similarly of a reference configuration, associated with an interval on a line,

$$0 \le x \le L, \tag{2.4.1}$$

the line now being normal to the plate in its reference configuration. Motion involves a displacement u, in a perpendicular direction, described by an equation of the form

$$u = u(x, t),\tag{2.4.2}$$

the analogue of the previous y. With it is associated the velocity,

$$\dot{u} = \frac{\partial u}{\partial t},\tag{2.4.3}$$

and the analogue of the stretch λ, the shear strain

$$\gamma = \frac{\partial u}{\partial x},\tag{2.4.4}$$

with $\gamma = 0$ in the reference configuration. There is a difference in that, physically, λ should be positive, while γ can be positive or negative. Associated with the shearing motion is a shear stress τ analogous to σ. The equations of motion are of the form

$$\rho\ddot{u} = \frac{\partial \tau}{\partial x} + f,\tag{2.4.5}$$

f acting in the direction of the displacement u or its opposite. Then, as analogues of (2.3.21), (2.3.24), and (2.3.28), we have relations of the form

$$\phi = \phi(\gamma, \theta),$$

$$\tau = \frac{\partial \phi}{\partial \gamma},\tag{2.4.6}$$

$$\eta = -\frac{\partial \phi}{\partial \theta},$$

or

$$\varepsilon = \varepsilon(\gamma, \eta),$$

$$\tau = \frac{\partial \varepsilon}{\partial \gamma},\tag{2.4.7}$$

$$\theta = \frac{\partial \varepsilon}{\partial \eta}.$$

For a temperature equation, we have the copy of (2.3.27),

$$\theta\dot{\eta} = \frac{\partial q}{\partial x} + r.\tag{2.4.8}$$

One difference is that here a notion of material symmetry is of some importance. For plates made of isotropic and some anisotropic materials, it is reasonable to assume that ϕ is an even function of γ,

$$\phi(\gamma, \theta) = \phi(-\gamma, \theta).\tag{2.4.9}$$

Reversing the sign of γ amounts to reversing the shear displacement. For, say, isotropic plates, one expects that this will be associated with forces of the same magnitude but opposite directions ($\tau \to -\tau$ and $f \to -f$), keeping $\theta(x,t)$ the same. So, (2.4.9) summarizes common experience of this kind. One can encounter cases where it is unreasonable to assume (2.4.9), in dealing with some kinds of crystal plates, for example. Later, (2.4.9) will play an important role in discussions of Martensitic transformations, phenomena which are rather common in crystalline solids.

As is rather obvious, various ideas and results used in analysing bars can be borrowed to analyze plates and vice versa, and we will not belabor the obvious.

2.5 Thermodynamic Experiments

Clearly, it is important to know the form of the function ϕ for particular bars or plates and there is a rather common strategy for using experiments to try to accomplish this. The same kinds of ideas are used in dealing with three-dimensional thermoelasticity theory. Later, we will consider some two- and three-dimensional theories. To be definite, we will discuss the bars. We would like to induce the bar to be in various configurations in which λ and θ take on a variety of values, independent of position. Physically, we would like to know that each of the pairs of values can be connected to the reference values ($\lambda = 1$, $\theta = \theta_R$) by a process which is, at least to a good approximation, reversible. Theoretically, entropy changes are involved in a rather important way and we need some assurance that it is physically well-defined.

Commonly, workers try three types of experiments:

1. measurements of thermal expansion;

2. measurements of specific heat $C_0(\theta)$ in unstressed specimens at various temperatures;

3. static mechanical experiments performed at constant temperature, at a variety of temperatures.

If all goes smoothly, this combination can, in principle, suffice and for a first look we will assume that it does.

Thermal expansion refers to the deformation caused by a change of temperature in an unloaded body. Theoretically, we would derive it by solving for λ,

$$\sigma = \frac{\partial \phi}{\partial \lambda}(\lambda, \theta) = 0, \tag{2.5.1}$$

to get

$$\lambda = \overline{\lambda}(\theta). \tag{2.5.2}$$

By assumption, we have chosen our reference configuration so that

$$\overline{\lambda}(\theta_R) = 1. \tag{2.5.3}$$

At this stage the function ϕ is unknown, but we can observe the stretch $\overline{\lambda}(\theta)$; it should be a uniform stretch, varying continuously with θ, for things to go smoothly. Also, as a check on reversibility, we should check that we get the same $\overline{\lambda}(\theta)$ if we increase or decrease θ. Often, thermal expansion is a small effect, so one finds some workers omitting this experiment. However, it is sensible to understand the scheme before considering such shortcuts.

Turning to the measurement of $C_0(\theta)$, we are concerned with changes in η occurring when we keep $\sigma = 0$ and change θ, hence with

$$\overline{\eta}(\theta) = \eta(\overline{\lambda}(\theta), \theta). \tag{2.5.4}$$

Conceptually, this is the same as measuring the specific heat for a rigid bar, with $\overline{\eta}(\theta)$ as its entropy function. By calorimetry, we can find

$$C_0(\theta) = \theta \frac{d\overline{\eta}}{d\theta} = \theta \left(\frac{\partial \eta}{\partial \theta} + \frac{\partial \eta}{\partial \lambda} \frac{d\overline{\lambda}}{d\theta} \right). \tag{2.5.5}$$

It is customary for workers making isothermal mechanical measurements normally to take as a reference configuration an unstressed configuration at the temperature θ_E at which the experiment is performed. Relative to our fixed reference configuration, the material point at x has moved to the position

$$y_E = \overline{\lambda}(\theta_E)x. \tag{2.5.6}$$

Adding loads then takes y_E to

$$y = \lambda_E y_E = \lambda_E \overline{\lambda}(\theta_E)x = \lambda x, \tag{2.5.7}$$

λ_E being what the experimentalist would term stretch. What he means by stress is force divided by area in some unstressed configuration. Suppose that this configuration is ours, that in the fixed reference configuration. If not, we need to know the relation between the areas in order to compare results which can involve another aspect of thermal expansion. Then, with standard simple tension and compression tests, the experimentalist seeks to determine relations of the form

$$\sigma = f(\lambda_E, \theta_E), \qquad f(1, \theta_E) = 0, \tag{2.5.8}$$

for some range of λ_E and θ_E. For all to go smoothly, σ should vary smoothly with λ_E and the same relation should hold whether we are increasing or decreasing σ. From this, we obtain the so-called isothermal strain energy function, given by

$$W(\lambda_E, \theta_E) = \int_1^{\lambda_E} f(\xi, \theta_E) \, d\xi, \tag{2.5.9}$$

with the properties that

$$\sigma = \frac{\partial W}{\partial \lambda_E} \tag{2.5.10}$$

and that is vanishes when $\lambda_E = 1$, that is, when $\sigma = 0$

$$W(1, \theta_E) = 0. \tag{2.5.11}$$

Given this, we can calculate the function representing energy per unit reference length,

$$\overline{W}(\lambda, \theta) = W \left(\frac{\lambda}{\overline{\lambda}(\theta)}, \theta \right) \overline{\lambda}(\theta), \tag{2.5.12}$$

where we have used (2.5.7). Then, by the chain rule,

$$\frac{\partial \overline{W}}{\partial \lambda} = \frac{\partial W}{\partial \lambda_E} \left(\frac{1}{\overline{\lambda}(\theta)} \right) \overline{\lambda}(\theta) = \frac{\partial W}{\partial \lambda_E} = \sigma. \tag{2.5.13}$$

Also, from (2.5.11) and (2.5.12), we have

$$\overline{W}(\overline{\lambda}, \theta) = 0. \tag{2.5.14}$$

Now, from (2.3.21) and (2.5.13), we should have

$$\sigma = \frac{\partial \overline{W}}{\partial \lambda} = \frac{\partial \phi}{\partial \lambda},$$

so

$$\phi = \overline{W}(\lambda, \theta) + f(\theta). \tag{2.5.15}$$

To determine $f(\theta)$ we note that, if we set $\lambda = \overline{\lambda}(\theta)$ and vary θ, $\overline{W}$ stays zero and, of course, $\sigma = 0$. Thus, bearing in mind (2.5.4), we have

$$d\phi = f'(\theta) \, d\theta = -\overline{\eta}(\theta) \, d\theta \quad \text{when } \sigma = 0. \tag{2.5.16}$$

So, given $C_0(\theta)$, from measurements of the specific heat at zero stress, we integrate (2.5.5) giving

$$\overline{\eta} = \int_{\theta_{\downarrow}}^{\theta} \frac{C_0(\xi)}{\xi} \, d\xi + a \tag{2.5.17}$$

with a an arbitrary constant. Then, using (2.5.16), we integrate once more, to get

$$f = - \int_{\theta_R}^{\theta} \int_{\theta_R}^{\upsilon} \frac{C_0(\xi)}{\xi} \, d\xi \, d\upsilon - a\theta + b, \tag{2.5.18}$$

where b is another integration constant. Physically, it is unimportant what values we assign to a and b; it is only differences in entropy and energy which are physically significant.

Intuitively, it is clear that the program can fail for various reasons. Specimens may break, buckle, or melt, for example. One can encounter phase transitions where quantities of interest change in a discontinuous manner, thus causing difficulties. Later, we will discuss some ideas for coping with them.

It is worth bearing in mind that while the conventional program discussed involves relatively simple experiments, when they work, the program is not infallible. Particularly when some mysterious difficulty is encountered, it is worth considering what other kinds of experiments may be performed to help illumine the situation.

Clearly, similar ideas can be used for the plates. For these, if the material symmetry is such that (2.4.9) applies, it is somewhat likely that the analogue of thermal expansion will not occur because $\tau = 0$ forces γ to be zero. As will later become clear, this is not always the case. When (2.4.9) does not apply, the analogue of thermal expansion will most likely occur, changes in temperature producing shear when $\tau = 0$. In either case, one may well get changes in volume not describable by our one-dimensional theory.

Ideas similar to those used in this chapter have been adopted to construct a great variety of three-dimensional theories of materials too complicated to be discussed here. Often, experimentalists and others interested in materials science do use one-dimensional theories; we have simply looked at a few of these to illustrate common types of thermodynamic reasoning used in formulating them.

2.6   Exercises

For rough analyses of rubber bars, workers often use the so-called Neo-Hookean theory, for which ϕ is of the form

$$\phi = a\theta(\lambda^2 + 2/\lambda - 3) - b[\theta \ln(\theta/c) - \theta + c],$$

where a, b, and c are positive constants. For Exercises 2.1–2.5, use this form.

2.1. What does this form predict for thermal expansion?

2.2. What constitutive equation for σ is appropriate for isothermal processes ($\theta = $ const)?

2.3. What constitutive equation for σ is appropriate for isentropic processes ($\eta = $ const)?

2.4. For isentropic processes, what is the strain energy function?

2.5. From thermodynamic experiments on a material, we find that the constitutive equation fits fairly well for a choice of the adjustable constants, but needs some correction. The fit for specific heat is satisfactory. The strain energy function fits, if we replace λ by λ_E, and curve-fitting measurements of thermal expansion gives $\bar{\lambda}(\theta) = 1 + k(\theta - \theta_R)$ for the temperatures occurring in the experiments, k being a constant. Work out the corrected constitutive equation for ϕ, and determine how this affects η.

2.6. For bar theories, the quantity

$$\delta = \frac{\partial \eta}{\partial t} - \frac{\partial(q/\theta)}{\partial x} - r/\theta$$

is commonly considered to be a measure of dissipation. For our theory of rigid bars, with

$$q = k(\theta)\frac{\partial(q/\theta)}{\partial x}, \quad r = \alpha[\theta_B(t) - \vartheta], \quad k > 0, \ \alpha > 0,$$

what can you say about the possibility of solutions of the temperature equation (2.2.13) that involve no dissipation ($\delta = 0$)?

2.7. Consider the graph of a smooth function $y = y(x)$ and the tangent line to it at a point $(\bar{x}, \bar{y} = y(\bar{x}))$. Show that points in the x-y plane lie above this line whenever their coordinates (x, y) satisfy the inequality

$$y - \bar{y} - y'(\bar{x})(x - \bar{x}) > 0,$$

and below it when the opposite inequality holds.

3
Equilibrium Theory of Bars

3.1 Equilibrium of Bars Subjected to Dead Loads

In commonly used jargon, a dead loading device is one that applies a constant force to some part of a specimen while leaving the part free to move. For example, one might load a bar by hanging it, clamping the upper end so it cannot move, and attaching weights to the lower end. Static experiments of this general kind have been done for centuries on a great variety of solid materials as is discussed in a lengthy article by Bell [13]. Typically, the apparatus is designed so one can add a small weight, allow it to come to equilibrium, then add a little more weight, to determine experimentally how stress is related to deformation. This is a rather standard way of approximating the condition of dead loading on the lower end with tensile loadings. By incorporating levers, for example, one can also apply compressive loads. So, for bars, this is one way of attempting to perform isothermal mechanical measurements needed to determine $\phi(\lambda, \theta)$ experimentally. Naturally, one must exercise some judgment in deciding whether the observations are compatible with the theory considered, in this case our one-dimensional bar theory. Actually, such theory can handle some phenomena which may seem to be outside its range. Thus, it is worthwhile to try to obtain a better understanding as to what it can do, with an open mind.

We now return to the theory of thermodynamic stability discussed in Chapter 1. Roughly, our thermodynamic system consists of the bar and the loading device in contact with a heat bath at a constant temperature θ_B.

That is, the definition of thermodynamic equilibrium implies that the temperature of the bar must reduce to that of the heat bath, plus the conditions which could also be obtained from standard reasoning in mechanics as conditions for static equilibrium. We have here one of many examples reinforcing the view that, physically, the general theoretical ideas of thermodynamic equilibrium are sensible.

Also involved is the idea that such equilibria may or may not be stable. For them to be stable (metastable), they should correspond to absolute (relative) minimizers of E_χ. Most workers are happy to make an assumption that makes such analyses easier and is not in obvious disagreement with experience; namely that, for any value of λ, ε is a strictly convex function of η. That is, for any λ, η_1, and η_2,

$$\varepsilon(\lambda, \eta_1) - \varepsilon(\lambda, \eta_2) - \frac{\partial \varepsilon}{\partial \eta}(\lambda, \eta_2)(\eta_1 - \eta_2) \geq 0. \qquad (3.1.14)$$

The equality holds only if $\eta_2 = \eta_1$. Assuming this, suppose that we choose η_2 depending on λ, so that

$$\frac{\partial \varepsilon}{\partial \eta}(\lambda, \eta_2) = \theta_B. \qquad (3.1.15)$$

Then, the inequality gives

$$\varepsilon(\lambda, \eta_1) - \theta_B \eta_1 \geq \varepsilon(\lambda, \eta_2) - \theta_B \eta_2 = \phi(\lambda, \theta_B). \qquad (3.1.16)$$

Now, suppose that we were to find $y = \hat{y}(x)$ minimizing the potential

$$F = \int_0^L \phi(\lambda, \theta_B) \, dx - ky(L), \qquad (3.1.17)$$

so that

$$F \geq \hat{F} = \int_0^L \phi(\hat{\lambda}, \theta_B) \, dx - k\hat{y}(L), \qquad (3.1.18)$$

perhaps with the proviso that λ be close to $\hat{\lambda}$. Then, using (3.1.16), we will have for the same $y(x)$

$$E_\chi \geq F \geq \hat{F} = \hat{E}_\chi. \qquad (3.1.19)$$

The last equality follows from the fact that any minimizer will have $\delta E_\chi = 0$, implying that its temperature must be θ_B. So, many workers will use the Helmholtz free energy in place of the ballistic free energy in applying the stability criterion and it does make the analysis easier. As is rather obvious and easily verified, (3.1.13) also gives the conditions obtaining from $\delta F = 0$.

One test for a relative minimum is the second derivative test, often used in such contexts. Represent $y(x)$ as in (3.1.6) to get F as a function of μ, and calculate

$$\delta^2 F = F''(0) = \int_0^L \overline{\frac{\partial^2 \phi}{\partial \lambda^2}} \left(\frac{\partial \delta y}{\partial x} \right)^2 dx, \qquad (3.1.20)$$

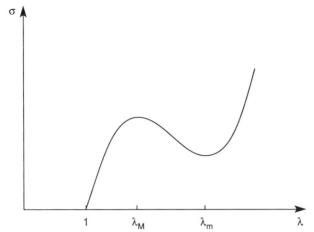

FIGURE 3.1. A simple nonmonotone stress-stretch curve, with horizontal tangents at λ_M and λ_m.

where, as before, the overbar denotes evaluation at the equilibrium state. For a relative minimum this should be nonnegative for any choice of δy such that $\delta y(0) = 0$. From this, it follows that we must have throughout the bar,

$$\overline{\frac{\delta^2 \phi}{\partial \lambda^2}}(\lambda, \theta_B) = \overline{\frac{d\sigma}{d\lambda}}(\lambda, \theta_B) \geq 0. \qquad (3.1.21)$$

Now, there are two possibilities. One is that the equation of state satisfies $\partial^2 \phi / \partial \lambda^2 > 0$ for all λ at this temperature, so σ is a monotonically increasing function of λ. Then, there is at most one value of $\lambda = \overline{\lambda}$ satisfying $\sigma(\overline{\lambda}, \theta_B) = k$, a given constant. Assuming there is one, then for this value, one gets the homogeneous deformation

$$y = \overline{\lambda} x,$$

satisfying (3.1.1). Also, one can show that it is the one and only minimizer of F. Physically, the theory is bound to fail for one reason or another if k gets too large. This calls for some judgment.

The other possibility is that the function $\sigma(\lambda, \theta_B)$ is not everywhere a monotonically increasing function of λ, a possibility which is seriously considered. A relatively simple possibility is to have a graph of the kind shown in Fig. 3.1.

With $\sigma = \partial \phi / \partial \lambda$, it is easy to see that the graph of ϕ at $\theta = \theta_B$ has the form shown in Fig. 3.2.

It has points of inflection at λ_m and λ_M and fails to be convex. Another possibility is a graph of $\sigma(\lambda, \theta_B)$ like that shown in Fig. 3.3, for which ϕ has a graph at θ_B diagrammed in Fig. 3.4, exhibiting a second minimum, which could be an absolute or relative minimum.

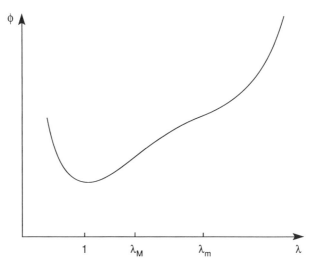

FIGURE 3.2. The graph of ϕ corresponding to Fig. 3.1, with points of inflection at λ_M and λ_m.

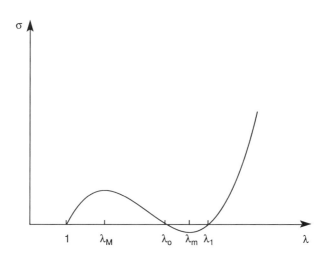

FIGURE 3.3. A stress-stretch curve similar to that in Fig. 3.1, but intersecting the σ-axis at three places, $\lambda = 1$, λ_0, and λ_1.

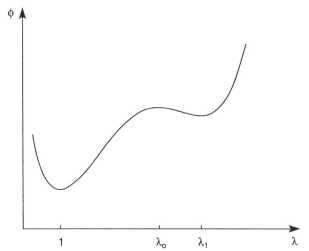

FIGURE 3.4. The graph ϕ corresponding to Fig. 3.3.

For either of these situations, one can find the absolute minimizers of F, the most stable configurations. With (3.1.1), we have

$$y(L) = \int_0^L \frac{\partial y}{\partial x}\, dx = \int_0^L \lambda\, dx, \qquad (3.1.22)$$

so we can write

$$F = \int_0^L [\phi(\lambda, \theta_B) - k\lambda]\, dx. \qquad (3.1.23)$$

Clearly, the integral will be smallest if the integrand is everywhere smallest which will occur at $\lambda = \overline{\lambda}$ such that

$$\phi(\lambda, \theta_B) - k\lambda \geq \phi(\overline{\lambda}, \theta_B) - k\overline{\lambda} \qquad (3.1.24)$$

for all λ. Of course, the first derivative test gives

$$\sigma(\overline{\lambda}, \theta_B) = \frac{\overline{\partial \phi}}{\partial \lambda} = k, \qquad (3.1.25)$$

where the overbar is interpreted as before. So, on the graph of ϕ we locate the points with slope k. Then, the inequality means that the graph of ϕ must be everywhere above the tangent line drawn through this point; we went through a mathematically identical consideration before in the considerations of rigid bars. Choose a constant $\overline{\lambda}$ satisfying these conditions, and $y = \overline{\lambda}x$ will be an admissible deformation minimizing F. Theoretically, this is the best that can be done in terms of stability. Somewhat stable (metastable) configurations are not ruled out by these considerations. In terms of a particular theory, the distinction between stable and metastable

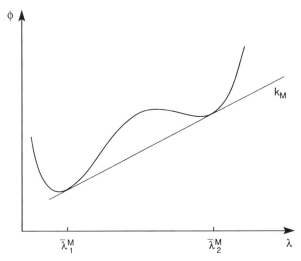

FIGURE 3.5. The common tangent line added to the graph in Fig. 3.4, k_M denoting its slope, $\bar{\lambda}_1^M$ and $\bar{\lambda}_2^M$ denoting values of λ at the points of tangency.

is as clear as the distinction between absolute and relative minima. However, physically the distinction is more artificial. For example, it is intuitively clear that compressive loads can cause a bar to buckle and some of the configurations we here call stable are, no doubt, physically unstable for this reason. To explore this, we would need a more general theory, capable of describing the buckled configurations and this is not the only possibility omitted from our considerations. More accurately, "stable" means the most stable of the possibilities we choose to consider and one needs to exercise good judgment in deciding which to consider.

Now, for graphs of ϕ of the kind indicated, there is a particular value of k for which two values of $\bar{\lambda}$ qualify in that they have a common tangent line below the graph, say at $k = k_M$.

Then,

$$\phi(\bar{\lambda}_1^M, \theta_B) - \phi(\bar{\lambda}_2^M, \theta_B) - k_M(\bar{\lambda}_1^M - \bar{\lambda}_2^M) = 0. \tag{3.1.26}$$

Now, for any value of k we know what it means for (3.1.24) and (3.1.25) to be satisfied by some value $\lambda = \bar{\lambda}$: the graph of $\phi(\lambda, \theta_B)$ must lie in the region above the tangent line, at $\lambda = \bar{\lambda}$. Geometrically, it is clear from the graph that this will occur for some $\bar{\lambda}$ such that

$$\bar{\lambda} < \bar{\lambda}_1^M \quad \text{when } k < k_M, \tag{3.1.27}$$

and

$$\bar{\lambda} > \bar{\lambda}_2^M \quad \text{when } k > k_M. \tag{3.1.28}$$

Thus except when $k = k_M$, the most stable configuration is obtained by setting

$$y = \bar{\lambda}x, \tag{3.1.29}$$

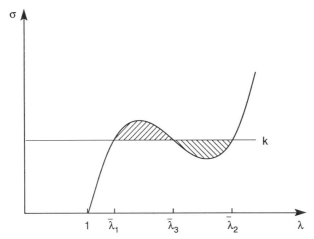

FIGURE 3.6. The stress-stretch curve in Fig. 3.1, hatchings indicating two areas associated with a load line $\sigma = k$.

where $\bar{\lambda}$ is the (unique) value of λ determined by these considerations. Clearly, this changes in a discontinuous manner as k passes through the value k_M.

There is another way of picturing the situation in terms of the graph of $\sigma(\lambda, \theta_B)$. We are concerned with values of k such that the line $\sigma = k$ intersects this graph in three places, as indicated by Fig. 3.6.

Let A_1 denote the hatched area indicated between $\bar{\lambda}_1$ and $\bar{\lambda}_3$. It is given by

$$A_1 = \int_{\bar{\lambda}_1}^{\bar{\lambda}_3} \sigma \, d\lambda - k(\bar{\lambda}_1 - \bar{\lambda}_3) = \phi(\bar{\lambda}_3, \theta_B) - \phi(\bar{\lambda}_1, \theta_B) - k(\bar{\lambda}_1 - \bar{\lambda}_3).$$

Similarly, the other hatched area, A_2, is given by

$$-A_2 = \phi(\bar{\lambda}_2, \theta_B) - \phi(\bar{\lambda}_3, \theta_B) - k(\bar{\lambda}_2 - \bar{\lambda}_3).$$

Thus,

$$A_1 - A_2 = \phi(\bar{\lambda}_2, \theta_B) - \phi(\bar{\lambda}_1, \theta_B) - k(\bar{\lambda}_2 - \bar{\lambda}_1). \qquad (3.1.30)$$

Comparing this with (3.1.26), we see that

$$A_1 = A_2 \quad \text{when} \quad k = k_M. \qquad (3.1.31)$$

This is the so-called *equal area rule*, first introduced by Maxwell in a mathematically similar analysis of van der Waals's fluids. It turns up in various physical problems. For $k < k_M$, one has $\bar{\lambda} = \bar{\lambda}_1$ and $A_1 > A_2$. For $k > k_M$, $\bar{\lambda} = \bar{\lambda}_2$ and $A_2 > A_1$.

We have presented reasoning that is very commonly accepted by workers interested in such phenomena. What is not widely appreciated is that it

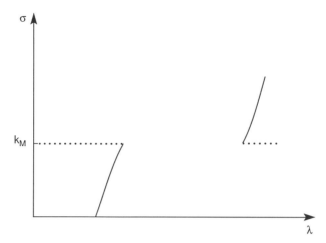

FIGURE 3.7. Loci of stable configurations, for the graph in Fig. 3.1, in a dead-loading device.

involves a rather subtle tacit assumption, pointed out by Kahl [14]. By modifying this one can come to different conclusions. We will ignore this complication, our aim being more to illustrate common practices.

Now, for the moment, suppose that the bar always chooses its most stable configuration as one increases k by small increments, starting from k = 0. It is then highly unlikely that one will hit the value k = k_M exactly. However, one will reach a point where adding a small load moves k through the critical value k_M, producing a sudden increase in the stretch, something many would regard as a phase transition. So, one will obtain data points like those shown in Fig. 3.7 and one should obtain essentially the same points when one similarly takes off weights for the thermoelastic model to apply, as we here interpret it. So, we simply do not get any equilibrium data points for an interval of values of λ. However, granted the reversibility, we can use (3.1.26) to estimate the change in ϕ associated with the transition which is helpful in the effort to determine this function experimentally. When the assumptions apply, the sudden jump provides an example of a process that is reversible but not reasonably considered to be quasi-static.

As a matter of experience, solids are rather prone to remain in metastable configurations, so the assumption just made may well fail to apply. Then, motivated to some degree by dynamical considerations discussed in Chapter 1, we accept one limitation on what can happen in time:

> *RULE: With* k *and* θ_B *fixed,* E_χ *may change to a lower value but it cannot change to a higher value.*

From the second derivative test, we have a minimal requirement for stability: configurations involving values of λ such that $d\sigma/d\lambda < 0$ are too

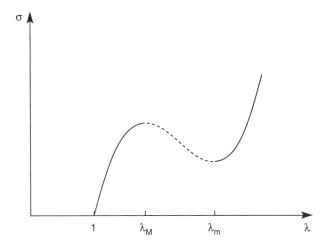

FIGURE 3.8. Loci of metastable configurations, for the graph in Fig. 3.1, in a dead-loading device.

unstable to be observed in the experiments considered. Indicating these by a broken curve, we have the effect shown in Fig. 3.8.

Following Gibbs, many refer to unstable regimes such as that occurring here when $d\sigma/d\lambda < 0$ as *spinoidal* regimes. Physically, stretches passing the second derivative test may or may not be stable enough to be observed. However, in loading up from $k = 0$, one is likely to stay on the left full curve in Fig. 3.9 until k becomes somewhat larger than k_M. Then there is a transition to the right curve, following this up as k increases, obtaining data points similar to those discussed before. If one then decreases k, as suggested in Fig. 3.9, the deformation cannot retrace its path, as this would violate the above rule. One must stay on the right branch at least until k reaches the value k_M. It may well stay here until k reaches a lower value, then shift to the more stable left branch. This gives rise to the phenomenon of hysteresis, with loading and unloading curves forming a loop like that shown in Fig. 3.9.

Rather commonly, one sees effects of this kind accompanying phase transition in solids. The values of σ at which transitions occur may or may not be very reproducible; they may be different for seemingly identical speci-mens loaded in the same device, for example. One sees a little more of the graph of $\sigma(\lambda, \theta_B)$ than previously. On the other hand, there is no good way to connect values of ϕ on the two branches from such data alone, since the change of ϕ accompanying a transition only satisfies an inequality, not an equation. By assumption, the left branch contains a point where $\sigma = 0$. It is possible that the right branch does also. If so, we might start on the left, load up enough to induce a transition, then unload, obtaining what seems to be permanent deformation. This particular process is not reversible. To justify the notion that our theory applies, one should try to find some other

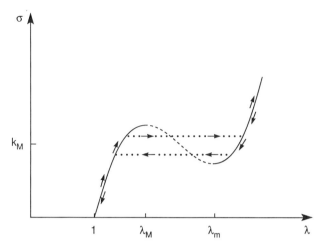

FIGURE 3.9. Hysteresis loop encountered when loads are increased, then decreased, involving metastable configurations. The arrows indicate where the load is being increased or decreased.

process connecting these two stretches, at $\theta = \theta_B$, which is, at least to a good approximation, reversible. In this, θ may well change as long as it begins and ends at the value θ_B.

In so-called semicrystalline polymers, one can sometimes obtain such processes by taking the polymer to a higher temperature, stretching it, and cooling it, holding the stretch fixed, removing the loads, then reversing this process. For these, heating the more stretched state when unloaded causes a sudden shift back to the reference configuration. The phenomenon is exploited in "heat-shrink" insulating material and in sealing up poultry in polymer coverings, for example. At least qualitatively, one can model such phenomena in the following way. Assume that, at a sufficiently high temperature, ϕ is a convex function of λ, but at lower temperatures it is more like that pictured in Fig. 3.4, having two minima, providing two unstressed configurations which are at least metastable. Behavior roughly like that predicted by such models is observed in various high polymers and is exploited in various applications. It seems to have first been exploited in making nylon fibers: inducing nylon to be in a more stretched state makes the fiber stronger. Some other kinds of models for describing such phenomena are discussed by different authors in reference [15], for example.

In this discussion we have assumed, rather tacitly, that at each value of k, λ is independent of x. Actually, there are possibilities for having λ take on different values in different parts of the bar. Shortly, we will be forced to consider this possibility.

Here, we have examples of the kinds of difficulties that can arise in trying to use experiments to determine $\phi(\lambda, \theta)$, associated with the occurrence of phase transitions. It is worth bearing in mind that it is not necessary to use only the standard kinds of experiments discussed earlier. An ingenious person knowing something about the difficulties encountered may find another way to obtain some valuable information.

Now, let us return to (3.1.25) and consider what happens when we change the ambient temperature θ_B a little, assuming that the bar continues to be in the stable configuration. Generally, the graph of ϕ versus λ will change a little since ϕ depends on θ. Physically, one will still get a transition, but k_M, $\overline{\lambda}_1^M$ and $\overline{\lambda}_2^M$ will become functions of θ_B, changing as θ_B does. Bearing this in mind, we can differentiate (3.1.26) with respect to θ_B, giving

$$0 = \left[\sigma\left(\overline{\lambda}_1^M, \theta_B\right) - k_M\right]\frac{d\overline{\lambda}_1^M}{d\theta_B} - \left[\sigma\left(\overline{\lambda}_2^M, \theta_B\right) - k_M\right]\frac{d\overline{\lambda}_2^M}{d\theta_B}$$
$$- \eta\left(\overline{\lambda}_1^M, \theta_B\right) + \eta\left(\overline{\lambda}_2^M, \theta_B\right) - \frac{dk_M}{d\theta_B}\left[\overline{\lambda}_1^M - \overline{\lambda}_2^M\right].$$

However, from the way things are defined, the two values of σ are both equal to k_M, so this simplifies to

$$\frac{dk_M}{d\theta_B} = -\frac{\eta\left(\overline{\lambda}_2^M, \theta_B\right) - \eta\left(\overline{\lambda}_1^M, \theta_B\right)}{\overline{\lambda}_2^M - \overline{\lambda}_1^M}. \tag{3.1.32}$$

Now, write

$$\Delta\eta = \eta\left(\overline{\lambda}_2^M, \theta_B\right) - \eta\left(\overline{\lambda}_1^M, \theta_B\right) = Q_L/\theta_B. \tag{3.1.33}$$

Here Q_L, called the *latent heat*, is the heat supplied to the bar when the stretch changes from $\overline{\lambda}_1^M$ to $\overline{\lambda}_2^M$ in a reversible process. It is not easy to measure with the loading device in the picture. However, if the bar always remains in the most stable configuration the transition should be close to reversible, if we load in order that it occurs for k very close to k_M. Then, by doing experiments at different temperatures, we can estimate the functions $k_M(\theta_B)$, $\overline{\lambda}_1^M(\theta_B)$ and $\overline{\lambda}_2^M(\theta_B)$ and calculate Q_L and $\Delta\eta$. When there is some hysteresis but fairly reproducible transition conditions, one can get two curves in the $\sigma - \theta$ plane, one indicating where $\overline{\lambda}_1$ transforms to $\overline{\lambda}_2$, the other where $\overline{\lambda}_2$ transforms to $\overline{\lambda}_1$. One finds workers using these to make educated guesses as to the quantities needed here. Of course, we can rewrite (3.1.32) and (3.1.33) in the form

$$\frac{d\sigma}{d\theta} = -\frac{Q_L}{\theta\Delta\lambda}, \tag{3.1.34}$$

where we have written σ in place of k_M and so on, to conform better to convention. This is sometimes called the *Clausius–Clapeyron equation*, the analogue of an equation with the same name which has long been used in connection with phase transitions in fluids. It is useful in connecting the values of η in the two phases. Clearly, one can do essentially the same analysis for the theory of plates mentioned earlier. We will discuss this later.

3.2   Equilibrium Theory of Bars in Hard Devices

In commonly used jargon, a hard device is a device sturdy enough to hold parts of a specimen in fixed positions. Various testing machines are designed to do this, at least to a good approximation. For our bars, the aim is to fix the end positions so we have, say,

$$y(0) = 0, \qquad y(L) = a > 0 \qquad (3.2.1)$$

where a is a parameter we can now control. As before, we assume no body forces act, $(f = 0)$ so, as long as a is fixed, no work is done on the bar. Therefore, it can be considered as a mechanically isolated system. As before, we consider it to be in contact with a heat bath at a constant temperature θ_B. From the discussion in Chapter 1, the situation is covered by statement III and we should use as thermodynamic potential the ballistic free energy indicated by (1.3.3), with E and S the energy and entropy of the bar, respectively. However, we follow the common practice of using instead the Helmholtz free energy,

$$F = \int_0^L \phi(\lambda, \theta_B) \, dx. \qquad (3.2.2)$$

This and the previous F are defined differently, but this should cause no confusion. Also, we will simplify notation by writing θ in place of θ_B. Of course, it is to be understood that all deformations allowed must satisfy (3.2.1), so λ must satisfy

$$a = y(L) = \int_0^L \lambda \, dx. \qquad (3.2.3)$$

To determine the possible equilibria, we proceed as before to get

$$\delta F = \int_0^L \frac{\overline{\partial \phi}}{\partial \lambda} \left(\frac{\partial \delta y}{\partial x} \right) dx = \int_0^L \overline{\sigma} \left(\frac{\partial \delta y}{\partial x} \right) dx$$

$$= -\int_0^L \frac{\overline{\partial \sigma}}{\partial x} \delta y \, dx + \overline{\sigma} \delta y \Big|_0^L \qquad (3.2.4)$$

$$= -\int_0^L \frac{\overline{\partial \sigma}}{\partial x} \delta y \, dx = 0,$$

the overbar denoting evaluation at $y = \overline{y}(x)$, the putative equilibrium state. The calculation presumes some smoothness of the functions involved, and we will need to reconsider this later. By reasoning now familiar, we can then conclude that

$$\frac{\overline{\partial \sigma}}{\partial x} = 0 \Rightarrow \overline{\sigma} = \mathrm{k} = \mathrm{const.}, \qquad (3.2.5)$$

but this constant is now **not** given. Similarly, one can, as before, use the second derivative test to get the condition

$$\frac{\overline{\partial^2 \phi}}{\partial \lambda^2} \geq 0 \tag{3.2.6}$$

as a minimal requirement for stability.

Now, if at the temperature considered ϕ is a convex function of λ, (3.2.6) will hold for any choice of λ. Then, with σ a monotonically increasing function of λ, (3.2.5) cannot hold for two different values of λ, so $\overline{\lambda}$ must be constant throughout the bar. Then, from (3.2.1), we must have

$$\overline{\lambda} = a/L, \qquad \overline{y} = ax/L. \tag{3.2.7}$$

Putting this value into the constitutive equation for σ, we can determine the value of the constant k. Or, if we measure $\overline{\lambda}$ and k, or the equivalent, we can use the information on such isothermal mechanical experiments to help determine the function $\phi(\lambda, \theta)$, as discussed before.

As is clear from our study relating to dead-load devices, (3.2.5) can be satisfied for more than one value of λ when ϕ is not a convex function of λ, at least for some values of k. Consider the possibility that just one value of λ occurs in the bar. Then (3.2.7) must hold and, for some values of a/L, (3.2.6) will fail to hold. Such configurations cannot even be metastable. So we must consider the possibility that $\overline{\lambda}$ is not constant throughout the bar. It is rather obvious that, for constitutive equations anything like those considered before, $\overline{\lambda}$ cannot vary smoothly with x and satisfy (3.2.5). For $\overline{y}(x)$ to be discontinuous would mean that the bar moves through itself, which is unreasonable, or that it breaks, a possibility which we will try to exclude, if we can. It will be considered later, in Chapter 7. Let us reconsider δF, allowing for the possibility that $\overline{\lambda}(x)$ and $\partial \delta y/\partial x$ are continuous except at $x = \overline{L} < L$, with δy remaining continuous at $x = \overline{L}$. Then, we calculate that

$$
\begin{aligned}
\delta F &= \frac{d}{d\mu} \int_0^L \phi\left(\overline{\lambda} + \mu \frac{\partial \delta y}{\partial x}, \theta\right) dx \bigg|_{\mu=0} \\
&= \frac{d}{d\mu} \left[\int_0^L \phi\left(\overline{\lambda} + \mu \frac{\partial \delta y}{\partial x}, \theta\right) dx + \int_{\overline{L}}^L \phi\left(\overline{\lambda} - \mu \frac{\partial \delta y}{\partial x}, \theta\right) dx \right]\bigg|_{\mu=0} \\
&= -\int_0^L \frac{\partial \overline{\sigma}}{\partial x} \delta y \, dx - \int_{\overline{L}}^L \frac{\partial \overline{\sigma}}{\partial x} \delta y \, dx - (\overline{\sigma}^+ - \overline{\sigma}^-)\delta y(\overline{L}) = 0,
\end{aligned}
$$

where $\overline{\sigma}^+$ is the limit approached by $\overline{\sigma}$ as $x \to \overline{L}$, with $x > \overline{L}$, and $\overline{\sigma}$ is the limit approached from the other side. As before, we can argue that (3.2.5) must hold in each of the two intervals by considering δy vanishing at the end points, so $\overline{\sigma} = k^+$ in one interval, $\overline{\sigma} = k^-$ in the other. Putting this back in δF, we see that we must have $k^+ = k^-$, so (3.2.5) must hold throughout

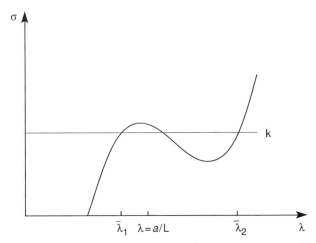

FIGURE 3.10. With the graph in Fig. 3.1, indicating quantities of interest when the overall stretch $\lambda = a/L$ falls in an unstable range.

the bar. Similar arguments apply if such discontinuities occur at more than one place. Also, by rather similar arguments, (3.2.6) must hold where $\bar{\lambda}$ is continuous. It is rather obvious that these conditions should hold for the forces to remain balanced. From this viewpoint, we are checking that the definitions of thermodynamic equilibrium are consistent with ideas of mechanical equilibrium. Consider a graph of $\sigma(\lambda, \theta)$ as we considered before and a value a/L in the range of concern, as shown in Fig. 3.10.

Although we do not know the value of k, we cannot succeed unless the line $\sigma = $ k gives us two values of $\bar{\lambda}$, $\bar{\lambda}_1$, and $\bar{\lambda}_2$ in the sketch, both satisfying (3.2.6). If it does, we can put the stretch equal to $\bar{\lambda}_1$ in some part, $\bar{\lambda}_2$ in the other, arranging that $\bar{y}$ is continuous. For example, for some $\bar{L} = bL$, $0 \leq b \leq 1$, set

$$y = \bar{\lambda}_1 x, \quad 0 \leq x \leq bL,$$
$$\bar{y} = \bar{\lambda}_2(x - bL) + \bar{\lambda}_1 bL, \quad bL \leq x \leq L. \tag{3.2.8}$$

Then, (3.2.3) gives

$$a = [\bar{\lambda}_1 b + \bar{\lambda}_2(1 - b)]L, \tag{3.2.9}$$

from which

$$b = \frac{(\bar{\lambda}_2 - a/L)}{(\bar{\lambda}_2 - \bar{\lambda}_1)}. \tag{3.2.10}$$

We want to ascertain that b lies between 0 and 1, which will be true if

$$\bar{\lambda}_1 \leq a/L \leq \bar{\lambda}_2. \tag{3.2.11}$$

From the figure, this is clearly true if $\lambda = a/L$ lies in the interval where $\partial^2 \phi / \partial \lambda^2 < 0$. However, such configurations are possible when it does not.

In the graph of σ they are possible as long as

$$\lambda_1^c \le a/L \le \lambda_2^c, \qquad (3.2.12)$$

with λ_1^c and λ_2^c denoting places where the horizontal tangents to the graph again intersect the graph, so this whole range is worth considering. For those configurations, it is easy to use (3.2.2) to calculate that

$$F = L[\phi(\lambda_1, \theta)b + \phi(\overline{\lambda}_2, \theta)(1 - b)]. \qquad (3.2.13)$$

Of course, the most stable of these will be those for which F has the smallest value, for θ and a/L fixed. Here, we have our first example of the possibility of an end-point minimum. Physically, b and $1 - b$ cannot take on negative values. Possibly, a configuration with $b = 1$ minimizes F, and we then cannot allow variations increasing b. Similarly, variations in b cannot decrease b, when $b = 0$. When the inequalities in (3.2.12) are strict, it is easy to see that we can get configurations of the kind considered for values of k in an interval, so we can vary this, producing variations in $\overline{\lambda}_1$, $\overline{\lambda}_2$, and b. From (3.2.13), we have, for the differential of F

$$dF = L\left[b\,\frac{\partial\phi}{\partial\overline{\lambda}_1}\,d\overline{\lambda}_1 + \phi(\overline{\lambda}_1,\theta)\,db + (1-b)\,\frac{\partial\phi}{\partial\overline{\lambda}_2}\,d\overline{\lambda}_2 - \phi(\overline{\lambda}_2,\theta)\,db\right]. \quad (3.2.14)$$

We already know that, for some value of k, we must have

$$\frac{\partial\phi}{\partial\overline{\lambda}_1} = \frac{\partial\phi}{\partial\overline{\lambda}_2} = k,$$

and, from (3.2.9), we must have

$$b\,d\overline{\lambda}_1 + (1 - b)\,d\overline{\lambda}_2 = (\overline{\lambda}_2 - \overline{\lambda}_1)\,db.$$

Using these, we can simplify (3.2.14), to get the condition for a minimum as

$$dF = L[\phi(\overline{\lambda}_1,\theta) - \phi(\overline{\lambda}_2,\theta) - k(\overline{\lambda}_1 - \overline{\lambda}_2)]\,db \ge 0. \qquad (3.2.15)$$

There are then three possibilities. For $b = 0$ to qualify, we can only have $db \ge 0$, so we might have

$$b = 0, \qquad \phi(\overline{\lambda}_1,\theta) - \phi(\overline{\lambda}_2,\theta) - k(\overline{\lambda}_1 - \overline{\lambda}_2) \ge 0. \qquad (3.2.16)$$

Similarly, we might have

$$b = 1, \qquad \phi(\overline{\lambda}_1,\theta) - \phi(\overline{\lambda}_2,\theta) - k(\overline{\lambda}_1 - \overline{\lambda}_2) \le 0. \qquad (3.2.17)$$

Finally, for other values of b, db can be positive or negative, giving

$$0 < b < 1, \qquad \phi(\overline{\lambda}_1,\theta) - \phi(\overline{\lambda}_2,\theta) - k(\overline{\lambda}_1 - \overline{\lambda}_2) = 0 \qquad (3.2.18)$$

and, from (3.1.24), this puts us on the Maxwell line, with

$$k = k_M, \qquad \bar{\lambda}_1 = \bar{\lambda}_1^M, \qquad \bar{\lambda}_2 = \bar{\lambda}_2^M,$$

this being determined by the equal area rule. In these terms, (3.2.16) translates to

$$b = 0, \qquad A_1 < A_2, \tag{3.2.19}$$

and (3.2.17) becomes

$$b = 1, \qquad A_1 > A_2, \tag{3.2.20}$$

with A_1 and A_2 the areas discussed in Section 3.1. With these results, it is easy to determine that the most stable configurations are given by

$$
\begin{aligned}
b &= 0 \qquad \text{when } \frac{a}{L} > \bar{\lambda}_2^M, \\
b &= 1 \qquad \text{when } \frac{a}{L} < \bar{\lambda}_1^M,
\end{aligned}
\tag{3.2.21}
$$

b is given by (3.2.10) with $\bar{\lambda}_1 = \bar{\lambda}_1^M$, $\bar{\lambda}_2 = \bar{\lambda}_2^M$ when $\bar{\lambda}_1^M \leq \frac{a}{L} \leq \bar{\lambda}_2^M$. So, again, the Maxwell line plays an important role. The most stable configurations are thus as indicated by Fig. 3.5. If the bar always chooses its most stable configurations, it should follow this path whether a is being increased or decreased, the stretch becoming discontinuous, to attain the desired overall stretch when necessary.

Again, solids are rather prone to "hang up" in metastable configurations and we may well get one to take on a variety of different metastable configurations depending on what we have done to it. Changing b slightly only changes F slightly. However, it causes some little part of the material to change its state quite significantly. If the two states are separated by a high energy barrier, such interfaces may well not move, leaving us with configurations that are really rather stable, although the energy would decrease if the interface would move. For example, we might attain a stable configuration of the kind in (3.2.21) involving a phase mixture. If we hold a fixed and change θ a little, one can see that b should change to adjust to the new Maxwell line, but experience indicates that it might not. Any criterion stronger than (3.2.6) is likely to exclude such possibilities. Again, there is an accepted rule, governing what can happen, similar to that discussed in Section 3.1.

> RULE: With θ and a fixed, it is possible that the bar's configuration will change to another if, and only if, the newer configuration has a smaller value of F.

For example, as we start increasing a from an initially unstressed, homogeneous configuration, it is not unlikely to continue to deform homogeneously until a/L becomes somewhat larger than $\bar{\lambda}_1^M$. The corresponding

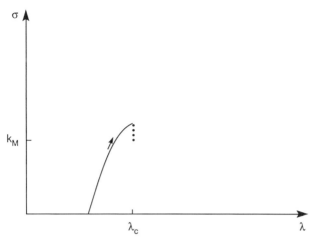

FIGURE 3.11. As the bar shifts from a metastable configuration, in a hard device, stress generally jumps, to the Maxwell line, with $\sigma = k_M$.

stress will then get larger than k_M. A reasonable possibility is that, at some point, it will change to the most stable configuration. When it does, the stress will drop to the value k_M and, from the rule, this process is not reversible. What we then see is a path of the kind shown in Fig. 3.11.

Often, one does see a drop in stress more or less like this as the "phase mixture" first forms. To try to decide whether our theory may continue to apply, we would like to know whether there is another way of moving from the initial state to this end state by a reversible process. This could involve intermediate changes of temperature, and so on. This kind of problem is not routine, so an experimentalist probing it needs to exercise some ingenuity. As one increases a from this point, one might stay on the Maxwell line until a/L reaches the value $\overline{\lambda}_2^M$. However, the rule does not exclude the possibility that a change of a again puts the bar in another metastable configuration, perhaps followed by another drop back to the Maxwell line. Said differently, equilibrium theory does not supply very definite predictions about metastable configurations. Other kinds of theory, for example, dynamical theory, sometimes do better, but our ability to make reliable predictions about such phenomena is very limited. In any event, it is clear that, theoretically, the behavior of bars in hard and soft devices can be rather different, and this is consistent with experience. Also, we have learned that there are rather subtle differences between the function $\sigma(\lambda, \theta)$ and the relations between σ and λ which may be observed in experiment. Phenomena of this kind do complicate the experimental program discussed in Section 2.5 but, to some degree, we can cope with them.

3.3 Exercises

3.1. Consider the constitutive equation for rubber used in Exercise 2.1–2.5. Suppose that we start with the bar in equilibrium, with $\lambda = 1$, $\theta = \theta_0$, then cause it to deform in an isentropic process, to get σ to be a constant tensile stress. Derive a formula for the corresponding change in temperature and determine whether the bar becomes hotter or colder. Roughly, this describes what should happen if the bar is stretched quite rapidly. Quickly stretch a rubber band and touch it to your lip and see if you experience what you predicted.

3.2. For the same material, we now clamp one end and hang a weight on the other, to dead load it, at room temperature, and let it come to equilibrium. Then we increase the ambient temperature and let it come to equilibrium at this temperature, without changing the load. Theoretically, should this heating produce an increase or decrease in the length of the bar? Try to design and carry out a simple experiment on a rubber band to check your prediction.

3.3. For a smooth function $y(x)$, defined for all x, show that the following two conditions are equivalent:

(a) *For all x, $y''(x) \geq 0$.*

(b) *For all choices of x_1 and x_2, $y(x_1) - y(x_2) - y'(x_2)(x_1 - x_2) \geq 0$.*

For Exercises 3.4–3.7, suppose that, at a temperature taken as the reference for some bars, the strain energy functions for this all have the form

$$W(\lambda) = \frac{a(\lambda - 1)^2}{2} + \frac{b(\lambda - 1)^3}{3} + \frac{c(\lambda - 1)^4}{4},$$

where a, b, and c are constants. To simplify considerations, you can assume that this applies for $-\infty < \lambda < \infty$, although this is physically unrealistic.

3.4. For what values of the constants is $W(\lambda)$ a convex function?

3.5. For what values of the constants does $W(\lambda)$ fail to be convex, but have the property that $\sigma = 0$ only when $\lambda = 1$?

3.6. For what values of the constants are there two values of λ for which $\sigma = 0$, both passing the second derivative test for stability in a deal-loading device at zero load?

3.7. For what values of the constants are there two values of λ for which $\sigma = 0$, both being stable equilibria, when the bar is considered to be dead-loaded?

4
Equilibrium Theory of Plates

4.1 Martensitic Transformations

The previous theory can be adapted, in an obvious way, to the one-dimensional theory of shear in plates. However, the latter does better than the former in illustrating some phenomena which are rather common in crystals, called Martensitic transformations. Here special kinds of material symmetry are important. With our plates, we can have this described by

$$\phi(\gamma, \theta) = \phi(-\gamma, \theta), \qquad (4.1.1)$$

as was discussed before. In this context, two pictures serve as representative. For the first, the graph of ϕ versus γ changes with θ as indicated in Fig. 4.1.

That is, it is convex at higher temperatures, with a minimum at $\gamma = 0$, this being replaced by minima at $\gamma = \pm\gamma_0$ when $\theta < \theta_c$, the value of γ_0 depending on θ, and approaching zero as $\theta \to \theta_c$. Of course, it is the material symmetry that forces ϕ to have a minimum at $-\gamma_0$ if it has one at γ_0. For $\theta > \theta_c$, the minimum at $\gamma = 0$ is associated with a phase called Austenite. The minima at $\gamma = \pm\gamma_0 \neq 0$ for $\theta < \theta_c$ are associated with "twin" phases, called *Martensite*. In real crystals, Austenite is, typically, a crystal configuration with greater symmetry than the Martensite phases. The Martensite phases have comparable symmetry. For example, one might be a mirror image of the other. Here, the minimizing values of γ depend continuously on θ. If measurements indicate this and there is no latent heat associated with the transition at $\theta = \theta_c$, these are what workers are likely to call second-order phase transitions. The other picture has the graphs changing as indicated in Fig. 4.2.

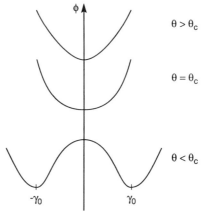

FIGURE 4.1. Sketch showing the shape of the graph of ϕ vs. γ for a typical second-order transition, at temperatures near the transition temperature θ_c, where $\frac{\partial^2 \phi}{\partial \gamma^2} = 0$ at $\gamma = 0$. The line indicates where $\gamma = 0$. The curves have been displaced from each other vertically, to make the picture clearer.

Again $\gamma = 0$ is associated with Austenite, $\pm\gamma_0$ with Martensite. For θ near θ_c, we have minima corresponding to Austenite and Martensite, one or the other being a relative minimum depending on the value of θ. Here, θ_c denotes the critical value of θ at which ϕ has three absolute minima. As before, γ_0 depends on θ, representing the analogue of thermal expansion in bars. However, unlike the previous case, we never have $\gamma_0 = 0$. As we reduce θ through the value θ_c the absolute minimum changes in a discontinuous manner, from $\gamma = 0$ to $\gamma = \pm\gamma_0$ and, typically, there will be some latent heat associated with this. Here, thermal expansion occurs in a dramatic way. This is a rather typical picture of what workers would call a first-order Martensitic transformation. These are more common than those of second-order. Martensitic transformations can involve various kinds of changes in crystal symmetry. Workers often use the theory of Landau [16] to estimate whether a particular kind of symmetry change is likely to occur as a second-order transition.

Originally, such transitions were discovered in steels, associated with changes of composition. One old way of making steel involves putting some carbon into iron, a cubic crystal. If the carbon content is low enough, one still gets cubic crystals, but when the content is sufficiently high, one gets tetragonal crystals. To analyze this, one needs to introduce additional variables to describe the amounts of the different ingredients. After such revision, one again uses the general ideas of Gibbs to formulate criteria for stability of equilibrium, as will be discussed later.

Now, for the problem at hand, stability of the plate under dead loading can be formulated by copying the anlysis for bars. Consider one face to be

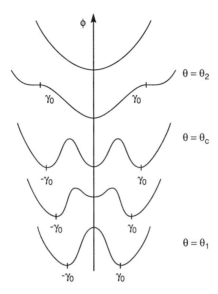

FIGURE 4.2. Sketch analogous to Fig. 4.1, for a typical first-order transition. Here, θ_c is the value of θ at which ϕ takes on its minimum value at three places. Upon cooling from higher temperatures, the graph first acquires points of inflection at $\theta = \theta_2$, then gets relative minima at $\gamma = \pm\gamma_0$, which become absolute minima for $\theta < \theta_c$. At $\theta = \theta_1$, the relative minimum at $\gamma = 0$ disappears, to become a relative maximum at lower temperatures. Bear in mind that γ_0 has different values at different temperatures, as suggested by the sketch.

held fixed, giving the boundary condition

$$u(0) = 0. \tag{4.1.2}$$

The other face is to be subjected to a constant shear force per unit area, k denoting its magnitude. As before, we have the arrangement in contact with a heat bath and, taking the usual short-cut, we consider this to be the temperature of the plate. Then, from the analogue of (3.1.17), we have

$$F = \int_0^L \phi(\gamma, \theta)\, dx - ku(L) \tag{4.1.3}$$

as the thermodynamic potential, whose absolute or relative minima correspond to stable or metastable equilibria. Recall that the shear stress τ and entropy density η are given by

$$\tau = \frac{\partial \phi}{\partial \gamma}, \qquad \eta = -\frac{\partial \phi}{\partial \theta}. \tag{4.1.4}$$

We might reconsider one point learned from our study of bars in hard devices. It may be reasonable to allow for the possibility that $\gamma = \partial u/\partial x$

could be discontinuous but, for the same reasons as before, we want $u(x)$ to be continuous.

As before, we get, as a condition for equilibrium, that

$$\tau = \frac{\partial \phi}{\partial \gamma} = \text{k}. \tag{4.1.5}$$

Here, we have not added overbars to denote evaluation at the equilibrium state: it should be sufficiently familiar to make this unnecessary. Similarly, at least where γ is continuous, the second derivative test for a minimum applies, giving

$$\frac{\partial^2 \phi}{\partial \gamma^2} = \frac{\partial \tau}{\partial \gamma} \geq 0. \tag{4.1.6}$$

Also, we can borrow the previous geometrical considerations of stability relating to the graph of ϕ versus γ. That is, on this graph, we find the tangent line(s) with slope k. If, at such a point, the entire graph lies in the half-plane above the line, the corresponding value(s) of γ generate stable configurations. For metastable configurations, only points on the graph near the point in question need lie in this half-plane. At high temperatures, where ϕ is a convex function of γ, the equilibria are unique and stable.

In terms of the graphs of ϕ, the geometrical picture of stability is similar to that for the bars. Given k, locate the point(s) where the tangent line(s) have this slope. For the corresponding values of γ, we can divide them into three sets. Unstable equilibria are of no interest. This includes values of γ for which $\partial^2 \phi / \partial \gamma^2 < 0$, or some with $\partial^2 \phi / \partial \gamma^2 = 0$. What is really relevant is whether any nearby points on the graph lie below the tangent line. If there are such points arbitrarily close, they correspond to unstable equilibria. If the whole graph lies above the tangent line, the corresponding value of γ can occur in a stable configuration. When there is more than one stable configuration, one can have either the homogeneous configurations with γ equal to one of these, or a phase mixture when γ takes on the different values in different parts of the plate. For metastable equilibria, it is still important that points on the graph lie above the tangent line, but only that part of it which is associated with values of γ sufficiently close to the value considered. Observations indicate that phase mixtures involving such metastable values of γ do occur.

For example, for ϕ behaving as indicated in Fig. 4.1, the unloaded configuration has $\gamma = 0$ as the only equilibrium configuration for $\theta \geq \theta_c$ and it is stable. For $\theta < \theta_c$ this is unstable but $\pm \gamma_0$ both qualify as stable. One can then construct an infinite number of stable phase mixtures with $\gamma = \gamma_0$ in some subintervals, $\gamma = -\gamma_0$ in the remainder. For rather obvious reasons the two phases are sometimes called twins, the mixture forming twinned Martensite. This can and should be done in such a way that $u(x)$

is continuous. For example, we could take

$$u(x) = \begin{cases} -\gamma_0 x, & 0 \le x \le L_1, \\ \gamma_0(x - L_1) - \gamma_0 L_1, & L_1 \le x \le L_2, \\ -\gamma_0(x - L_2) + \gamma_0(L_2 - 2L_1), & L_2 \le x \le L. \end{cases} \qquad (4.1.7)$$

where L_1 and L_2 are any numbers satisfying

$$0 < L_1 < L_2 < L.$$

Such configurations often occur spontaneously as one cools Austenite to obtain Martensite. The second sequence of pictures, in Fig. 4.2, becomes more complicated for θ near θ_c where both Austenite and Martensite are at least metastable and, at $\theta = \theta_c$, both are stable. Theoretically, one can then construct phase mixtures involving Austenite and twinned Martensite, which are metastable near $\theta = \theta_c$ and stable at $\theta = \theta_c$. In real crystals, one does see Austenite and twinned Martensite occurring in the same specimen, as suggested by this. Often, one sees regions where the twins occur as parallel planes, as suggested by one-dimensional theory. However, the interfaces between Austenite and Martensite generally have quite a different direction, so it is necessary to use three-dimensional theory to do realistic analyses of these.

There is another matter which is one of some interest. Suppose that one has twinned Martensite and a shear stress is applied to it as a small dead load, with k > 0, say. On the graph of ϕ, this will give tangent lines with this slope at values of γ, one near γ_0, the other near $-\gamma_0$. It is easy to see that, for the one near $-\gamma_0$, the graph of ϕ lies partly below the tangent line, so this is now only metastable. Similarly, one sees that the one near γ_0 remains stable. If one increases k sufficiently, that closer to $-\gamma_0$ becomes unstable, so, for some intermediate value of k, the twinned configuration must shift to the more stable homogeneous configuration. Mechanical treatments like this are sometimes used to eliminate twins which can be undesirable for some applications. This trick was discovered during World War II. The Allies, needing quartz for piezoelectric crystals, were forced to use what occurred naturally in France. This contained so-called Dauphiné twins which affected the performance of the crystals. This is not really quite what experts mean by twinned Martensite, but the basic idea is similar. With some rather elementary theory, eventually published by L.A. Thomas and W.A. Wooster [17] they worked out schemes for loading to maximize the energy differences which were used successfully to obtain better crystals. Later, similar treatments were, and still are, used on various other kinds of twinned crystals. The experience is that it does not work if the loads are too small, again indicating some tendency for such phase mixtures to "hang up" in metastable configurations. So, one makes the loads exceed a critical value, determined empirically. There is no reliable theory for determining this, as far as I know.

Later, various workers became interested in alloys exhibiting shape memory effects. These contain phase mixtures of Austenite and Martensite, with a morphology easily changed in a rather reversible way by the application of loads or temperature changes. Wrap a wire around your finger and it will seem to deform permanently, as would a soft copper wire. However, unlike the latter, one can get it to spring back to its original form by raising the temperature sufficiently. Solid-state engines have been made which involve alternating the temperature between values above and below the transition temperature and various other kinds of devices are being developed by entrepreneurs. The transformations do bear some resemblance to what occurs in steels, for example, but in the latter, processes changing the morphology tend to be far from reversible. With the structural metals, the metallurgist tries to "lock in" morphologies, to attain desired features of strength. A great deal of information concerning such materials is available in conference proceedings, [18–19]. Much information on Martensitic transformations in other materials is presented by Nishiyama [20]. Related three-dimensional mathematical theory is discussed by James and Kinderlehrer [21], for example. Such theory is becoming more sophisticated, a trend that is likely to continue.

4.2   Bifurcation Diagrams

In considering transition phenomena, workers find useful another kind of picture, sometimes called a bifurcation diagram. Essentially, it is a picture of equilibria of interest, indicating how they vary and branch as the variables of interest change. For our Martensitic transformations, one point of interest is the behavior of unloaded configurations as we change θ. The possible equilibria are then described by

$$\tau(\theta, \gamma) = 0, \tag{4.2.1}$$

which we can picture as curves in the θ–γ plane. One of these is the trivial solution, giving us the straight line $\gamma = 0$.

For our typical second-order transition, there are no other points where $\tau = 0$ when $\theta > \theta_c$, but we get two more when $\theta < \theta_c$, corresponding to $\gamma^2 = \gamma_0^2(\theta)$, and γ_0^2 is small when θ is close to θ_c. For small γ, we can approximate ϕ by the first few terms in a Taylor expansion, if it is smooth enough,[1]

---

[1] Actually, experiments on "critical exponents" indicate thermodynamic potentials are likely to have rather mild singularities at values of their arguments at which second-order transitions occur, these being analogous to "critical points" in fluids. Some discussion of these matters is given by Sengers *et al.* [22].

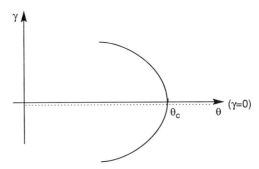

FIGURE 4.3. Graph of the curves $\tau = 0$, for a typical second-order Martensitic transformation, with θ_c as indicated in Fig. 4.1. The dotted line is to remind us that the line $\gamma = 0$ is included.

$$\phi \cong a(\theta) + \frac{b}{2}(\theta)\gamma^2 + \frac{c(\theta)}{4}\gamma^4$$

$$\tau \cong \gamma(b + c\gamma^2) \tag{4.2.2}$$

$$\frac{\partial^2\phi}{\partial\gamma^2} \cong b + 3c\gamma^2.$$

At $\gamma = 0$, we should have $b = \partial^2\phi/\partial\gamma^2$ changing from positive to negative as θ decreases and passes through θ_c. At $\theta = \theta_c$, ϕ has a minimum at $\gamma = 0$, indicating that $c(\theta_c) \geq 0$ and, as a general rule, $c(\theta_c) > 0$, so we assume this. Then, by continuity, $c(\theta_c) > 0$, for θ sufficiently close to θ_c. Setting $\tau = 0$, we then get for $\theta < \theta_c$ and $(\theta_c - \theta)$ small

$$\gamma_0^2 \approx -b(\theta)/c(\theta) > 0. \tag{4.2.3}$$

Also, in first approximation,

$$\gamma_0^2 \approx -b'(\theta_c)(\theta - \theta_c)/c(\theta_c), \tag{4.2.4}$$

indicating that this curve resembles a parabola near $\theta = \theta_c$, crossing the line $\gamma = 0$ with infinite slope. The curves $\tau = 0$ look rather like a pitchfork, as illustrated in Fig. 4.3.

"Pitch-fork bifurcations" giving pictures like this occur in a variety of physical situations. Later, we will encounter another associated with biaxial stretching of rubber sheets. Physically, for $\theta < \theta_c$, the line $\gamma = 0$ is too unstable to be observed, but it is a good idea to record such points in a bifurcation diagram. This can help to provide a better picture of the connection between branches.

With the typical first-order transition, we get, at some temperatures, five points where $\tau = 0$, three at the lowest temperatures. At lower temperatures there is a temperature θ at which the local minimum at $\gamma = 0$ changes to a local maximum, with $\partial\tau/\partial\gamma = 0$ at $\theta < \theta_1$, $\gamma = 0$. Very near this the

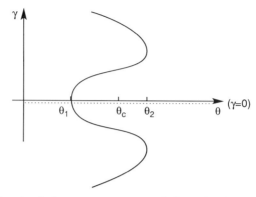

FIGURE 4.4. Graph of the curves $\tau = 0$, including the dotted line indicating inclusion of $\gamma = 0$, for a typical first-order Martensitic transformation, θ_1 and θ_2 being values indicated in Fig. 4.2.

picture looks like that above, with θ_c replaced by θ_1, except that the double well occurs at higher, instead of lower, temperatures as one can check by essentially the same analysis. At higher temperatures, $\tau = 0 \Rightarrow \gamma = 0$ for $\theta > \theta_2$, some value of θ, and we begin to get five points for $\theta < \theta_2$. On the right side of the graph are two close together, for $(\theta_2 - \theta)$ positive and small. As $\theta \to \theta_2$ they come together to give a single point (θ_2, γ_2), with

$$\tau = \frac{\partial \tau}{\partial \gamma} = 0 \quad \text{at } (\theta_2, \gamma_2), \tag{4.2.5}$$

as one can see from considerations of continuity. Now differentiating $\tau(\gamma, \theta) = 0$, we get the differential equation

$$\frac{d\theta}{d\gamma} = -\frac{\partial \tau / \partial \gamma}{\partial \tau / \partial \theta}, \tag{4.2.6}$$

giving a differential equation for a curve. Generally, $\partial \tau / \partial \theta$ will not vanish at (θ_2, γ_2), so the right side will be a smooth function of θ and γ. Solving this with the initial condition $\theta = \theta_2$ when $\gamma = \gamma_2$ then gives θ as a function of γ, with $d\theta/d\gamma = 0$ initially. This indicates that our two points at fixed θ, $\gamma > 0$ really lie on the same curve. To make a long story short, one gets the picture sketched in Fig. 4.4.

Pictures like this are also quite commonly encountered in physical problems and are sometimes said to involve subcritical bifurcations. Here, if we considered lowering the temperatures from high values and relied on second derivative tests for stability, we would see no indication of instability as long as $\theta > \theta_1$. Yet for any $\theta < \theta_c$, $\theta > \theta_1$, the system is only metastable and can be expected to transform before θ reaches θ_1. Bifurcation theory includes techniques for finding crossing branches but it is hard to locate isolated branches. Here, the branch crossing $\gamma = 0$ seems at first to be irrelevant. However, follow it and you will be led to the outer parts, which are, in fact, stable.

In dealing with near-transition phenomena, it is rather common to assume ϕ is a polynomial and use available experimental data to estimate the coefficients. For our second-order transitions, most would assume the quartic used above, with c a positive constant, b proportional to $(\theta - \theta_c)$, with $a(\theta)$ fit to available data on specific heats. For first-order phase transitions, γ^2 can be fairly small. Particularly in such cases, workers are likely to try assuming

$$\phi = A + \frac{B}{2} \gamma^2 + \frac{C}{4} \gamma^4 + \frac{D}{6} \gamma^6, \tag{4.2.7}$$

where the coefficients are simple functions of θ, again using available data to estimate these.

With suitable choices of these functions, one can obtain graphs of ϕ which change with θ in the manner indicated by the typical pictures. One needs a polynomial of degree six, at least, to provide five values of γ at which $\tau = 0$. Basically, it is for this reason that workers use a sextic, the polynomial of lowest degree which can describe the phenomena, at least qualitatively.

Roughly, a bifurcation diagram is a picture of the response of a system when a control variable is changed. Commonly, it is marked to distinguish the unstable from more stable parts. Here, we consider controlling θ, keeping $\tau = 0$. As another example, an isothermal stress–strain graph, suitably marked, could be viewed as a bifurcation diagram if we were controlling τ, keeping θ fixed.

4.3 Exercises

4.1. Suppose that, for a plate, ϕ is given by a constitutive equation of the form

$$\phi = \frac{(a\vartheta - b)\gamma^2}{2} + \frac{c\gamma^4}{4} + \frac{d\gamma^6}{6} + f(\theta),$$

where a, b, c, and d are constants, $f(\theta)$ being some smooth function. Show that this is the polynomial of lowest degree in γ which can describe a first-order Martensitic transformation.

4.2. For this form of constitutive equation, decide what are reasonable assumptions concerning algebraic signs of the constants to describe a first-order Martensitic transformation. Then derive formulae, expressing the following quantities in terms of these constants:

θ_1 The temperature at which Austenite changes from metastable to unstable.

θ_2 The upper bound of temperatures for which Martensite is metastable.

θ_c The temperature at which Martensite and Austenite are both stable.

$\gamma_0(\theta)$ The shear strain in unstressed Martensite.

4.3. For this example, decide what should be meant by the strain energy for the Martensite and give a formula for it.

4.4. For the special cases of the above constitutive equation with $d = 0$, which can describe second-order Martensitic transformations, calculate the corresponding $\gamma_0(\theta)$.

4.5. For the same special cases, at some temperature below θ_c, consider an unstressed sample containing a pair of twins with the discontinuity in the middle of the specimen. Then consider dead-loading the specimen a bit, with $\tau > 0$. Suppose that the discontinuity does not disappear or move. Find a formula for the difference in strain energies for the two parts as a function of τ. If necessary, you may use approximations appropriate for small τ.

4.6. Consider the analog for shear of the Clausius–Clapeyron equation (3.1.34).

$$\frac{d\tau}{d\theta} = -\frac{Q_L}{\theta\,\Delta\gamma},$$

Using results obtained in Exercise 4.2, calculate

$$\lim_{\theta\to\theta_c} \frac{d\tau}{d\theta}.$$

5
Balloon Problems

5.1 Equilibrium of Spherical Balloons at Fixed Pressure

Here, we will be concerned with rubber balloons subjected to inflating pressures. For simplicity, we assume they are spherical and remain so under such loads. It is known that rubber is, to a very good approximation, incompressible. Also, it can reasonably be considered to be homogeneous and isotropic.

As before, we introduce a reference configuration, an unloaded, spherical shape, at a reference temperature which we consider as room temperature. Material points will then lie on spheres of radius R, with

$$R_1 \leq R \leq R_2, \tag{5.1.1}$$

where R_1 and R_2 are, respectively, the inner and outer radii of the balloon in the reference configuration. A deformation then changes R to $r = r(R)$, with

$$r_1 = r(R_1) \leq r(R) \leq r(R_2) = r_2. \tag{5.1.2}$$

The incompressibility assumption means that the volume of that part between R_1 and R must be the same as that between r_1 and $r(R)$ or

$$r(R)^3 - r_1^3 = R^3 - R_1^3. \tag{5.1.3}$$

For a given balloon, $r(R)$ is thus determined by $r_1 = r(R_1)$. Also, for typical balloons,

$$(R_2 - R_1)/R_1 << 1,$$

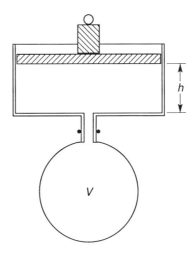

FIGURE 5.1. One device for controlling the pressure in a spherical balloon.

they being so thin that we may well think of them as infinitely thin in considering measurements of radius, which we shall do. Such thinking motivates us to assume that one number will suffice to describe the configurations considered. A dimensionless measure of stretch, similar to that used for bars, is

$$\lambda = r/R. \tag{5.1.4}$$

However, calculations become easier if we use instead a measure based on the volume V enclosed by the balloon

$$V/V_R = r^3 R^3 = \lambda^3. \tag{5.1.5}$$

Then, a likely assumption is that, for the balloon, the Helmholtz free energy function is of the form

$$F = F(V/V_R, \theta). \tag{5.1.6}$$

Then, we consider the possibility of using experiments to help determine the form of this function, in particular the isothermal mechanical experiments involved in a typical program of this kind. For simplicity, we think of working at a single temperature, room temperature, ignoring the dependence on θ.

We have in mind that no force will be applied to the exterior of the balloon, so this part will be mechanically isolated. The volume V will be filled with a gas to supply an inflating pressure p.

Let us first consider a device designed to try to control p. It involves a cylinder and a piston atop which we can add weights, connected by a small opening to the balloon, as indicated in Fig. 5.1.

Into this, we have put a measured mass M of gas. If A is the area of the piston, the total volume V_G occupied by the gas is given by

$$V_G = Ah + V, \tag{5.1.7}$$

neglecting the small amount in the connecting tube. For this, we do need some information, namely, an equation of state. For a gas, the likely models are of the form

$$F_G = M\phi\left(\frac{V_G}{M}, \theta\right), \tag{5.1.8}$$

the gas pressure being given by

$$p = -\frac{\partial F_G}{\partial V_G}, \tag{5.1.9}$$

with

$$\frac{\partial p}{\partial V_G} = -\frac{\partial^2 F_G}{\partial V_G^2} < 0. \tag{5.1.10}$$

A likely possibility is to use the ideal gas model with

$$p\frac{V_G}{M} = \theta \times \text{const.} = \text{const.} \tag{5.1.11}$$

at room temperature. With (5.1.9), this determines F_G to within a function of θ, which is good enough for our purposes. Except in consideration of the weight W on top, we ignore the effects of gravity. With it is associated the potential energy which can be taken as Wh. Then, adding up the three contributions, we get the thermodynamic potential

$$E = Wh + F_G + F. \tag{5.1.12}$$

Again, we have taken the usual shortcut, assuming that the temperature throughout is the ambient temperature. Here, the idea is that h and V can be varied independently, with the proviso that neither can become negative. Physically, there is a real possibility of having equilibria with the piston sitting on the bottom of the cylinder with $h = 0$. Physically, the possibility of having $V = 0$ is unreasonable. Then, possible variations in h cannot decrease it, this being like the end point minima discussed before. For the moment, let us exclude this possibility. Then, we obtain, as one equilibrium equation

$$\frac{\partial E}{\partial h} = 0 = W + \frac{\partial F_G}{\partial V_G}(A)$$

$$= W - pA, \tag{5.1.13}$$

or

$$p = W/A, \tag{5.1.14}$$

the rather obvious balancing of the weight with the gas pressure. It is in this sense that we control p and, clearly, it need not hold when $h = 0$. As the other equilibrium equation, we have, using (5.1.6),

$$\frac{\partial E}{\partial V} = 0 = \frac{\partial F_G}{\partial V_G} + \frac{\partial F}{\partial (V/V_R)} \cdot \frac{1}{V_R}, \tag{5.1.15}$$

or

$$p = \frac{\partial F}{\partial (V/V_R)} \cdot \frac{1}{V_R}. \tag{5.1.16}$$

Now, given (5.1.11) or another such equation satisfying (5.1.10), p is a monotonically decreasing function of V_G with M fixed, as it is here. Then we can solve (5.1.14) for V_G as a function of W/A. Note that this also gives F_G as a function of W/A. Here, experienced workers would be likely to take another shortcut. Take it as obvious from the start that (5.1.14) will hold. Then, for fixed p, we have

$$Wh = pAh = p(V_G - V) = -pV + \text{const.},$$

and F_G also becomes a constant, so we can replace (5.1.12) by the potential

$$\overline{E} = -pV + F, \tag{5.1.17}$$

(5.1.16) being the condition that $\partial \overline{E}/\partial V = 0$, $-pV$ being a kind of potential energy associated with work done on the balloon by the gas. If we use (5.1.16), the second derivative test for a minimum becomes simply

$$\frac{\partial^2 \overline{E}}{\partial V^2} = \frac{\partial^2 F}{\partial (V/V_R)^2} \frac{1}{V_R^2} \geq 0, \tag{5.1.18}$$

a test for metastability, p being now regarded as fixed.

Had we not taken the shortcut, we would be concerned with the second derivative test as it applies to $E(h, V)$, a function of two variables. For this, one calculates the second differential, and requires that this be nonnegative,

$$d^2 E = \frac{\partial^2 E}{\partial h^2} (dh)^2 + 2 \frac{\partial^2 E}{\partial h \partial V} dh\, dV + \frac{\partial^2 E}{\partial V^2} dV^2 \geq 0, \tag{5.1.19}$$

for all dh and dV. Conditions necessary and sufficient for this are that

$$\frac{\partial^2 E}{\partial h^2} \geq 0,$$

$$\frac{\partial^2 E}{\partial V^2} \geq 0, \tag{5.1.20}$$

$$\left(\frac{\partial^2 E}{\partial h \partial V} \right)^2 \leq \frac{\partial^2 E}{\partial h^2} \frac{\partial^2 E}{\partial V^2},$$

the second derivative test being inconclusive if any of the equalities hold. Using (5.1.10), one can show that this does, in fact, give the same condition as (5.1.18).

It remains to consider the possibility of equilibria with $h = 0$. Then, the differential E should be non-negative for possible dh and dV,

$$dE = \frac{\partial E}{\partial h}\, dh + \frac{\partial E}{\partial V}\, dV \geq 0. \tag{5.1.21}$$

Here, dh can be positive, but not negative, so we get

$$\frac{\partial E}{\partial h} = W - pA \geq 0,$$

or

$$p \leq W/A. \tag{5.1.22}$$

Also, since dV is not restricted, (5.1.16) still holds. Further, if equality holds in (5.1.22), $dE \equiv 0$ and one can still use (5.1.19) as a second derivative test. It is not hard to see that if $d^2 E \geq 0$ for all dV and all positive dh, it must also be nonnegative for negative dh so, again, (5.1.20) must hold.

If we use the simpler potential, a simple and now familiar picture gives the most stable configuration with $h > 0$. Rewrite (5.1.16) and (5.1.17) as

$$\bar{E} = F\left(\frac{V}{V_R}\right) - k\,\frac{V}{V_R}, \qquad k = pV_R = \frac{\partial F}{\partial(V/V_R)}, \tag{5.1.23}$$

k being a load parameter that is being controlled. On the graph of F versus V/V_R, locate the tangent lines with slope k, such that the entire graph lies above them. Any such configuration corresponds to a most stable configuration with $h > 0$.

For the configurations with $h = 0$, (5.1.7) implies that $V_G = V$. The most stable of these are then those that minimize

$$E\big|_{h=0} = F_G\left(\frac{V}{M}\right) + F\left(\frac{V}{V_R}\right), \tag{5.1.24}$$

with M fixed. We will discuss this more later. Such minimizers may be unstable to variations increasing h, for example, because (5.1.22) is violated, or because the second derivative test indicates this. In the end, it may be necessary to do a numerical calculation to determine whether the actual minimizer of E occurs with $h = 0$ or $h > 0$, even when the constitutive equations are known.

Observations of balloons indicate[1] that, at some values of pressure, we can have more than one value of V occurring in equilibrium configurations. If F always satisfies (5.1.18), it would be impossible for (5.1.15) to be satisfied by two values of V at a given value of p. Thus, it is rather clear that F is not a convex function of V/V_R. Suppose that we have a situation such as that previously considered for bars, involving a graph like that in Fig. 5.2.

---

[1]Simple experiments demonstrating this are discussed by Beatty [23] and Kitsche *et al.* [24].

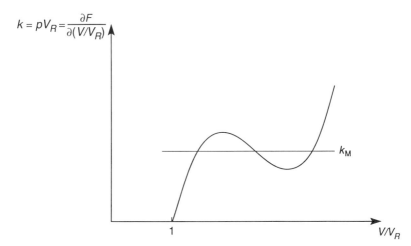

FIGURE 5.2. Graph indicating a reasonable kind of material response for a balloon.

For equilibria with $h > 0$, we can use (5.1.23), clearly analogous to that used for bars in soft loading devices. The equal area rule then gives us the Maxwell line. The downward sloping part of the graph corresponds to unstable equilibria. If we start increasing pressure in our device by increasing W, we know what we should see. That is, V should increase smoothly with pressure until we get to the Maxwell line, or a little higher if it tolerates being in a metastable configuration. Then, at some pressure, we should see values of V attained in equilibrium change in a discontinuous manner to find a more stable location on the part of the curve to the right. What is observed is that the transition occurs smoothly, but most rapidly, indicating that nonequilibrium theory would be needed to describe this. It occurs at a value of V/V_R close to $\sqrt{7}$, a value not much affected by the quality of the rubber, nor by modest departures from spherical shape, such as one gets in toy balloons which usually are not quite spherical. Of course, the possibility that we may attain equilibrium with $h = 0$ complicates the picture. If we ignore this, experiments of this kind involving measurements of p and V can give us only part of the curve. Observation of the jump in V would provide some support for the notion that the graph has a shape more or less similar to that illustrated.

5.2 Equilibrium of Balloons Containing a Fixed Mass of Gas

When we blow up a balloon, we insert some fixed mass M of gas into the balloon, then close off the opening. It is interesting to find out what we need for stability under these circumstances and it will help us to understand the previous situations better when $h = 0$. Here, only the balloon and gas are involved, and we will always have

$$V_G = V. \tag{5.2.1}$$

The relevant potential now is given by (5.1.12), which we write as

$$E = F_G \left(\frac{V}{M} \right) + F \left(\frac{V}{V_R} \right), \tag{5.2.2}$$

and we are interested in the absolute and relative minima of this, corresponding to stable or metastable equilibria. The first derivative test gives

$$\frac{\partial E}{\partial V} = 0 = \frac{\partial F_G}{\partial V} + \frac{\partial F}{\partial (V/V_R)} \frac{1}{V_R}, \tag{5.2.3}$$

or, as before,

$$p = \frac{\partial F}{\partial (V/V_R)} \frac{1}{V_R}. \tag{5.2.4}$$

However, the second derivative test now gives

$$\frac{\partial^2 E}{\partial V^2} = \frac{\partial^2 F_G}{\partial V^2} + \frac{\partial^2 F}{\partial (V/V_R)^2} \frac{1}{V_R^2} \geq 0. \tag{5.2.5}$$

From (5.1.10), the first term is positive; so, unlike before, we can have configurations which are at least metastable, when the second term is negative, but small enough to be dominated by the first term. Roughly, the pressure is now free to adjust and it works to help stabilize such configurations. Clearly (5.2.5) may hold for all or only part of the range where the second term is negative, depending on the detailed form of the equations. Experiments on balloons indicate that one can cover all, or at least a good part of the apparently unstable range with this kind of experiment. So, for determining the function F, this type of experiment is better than that considered before.

Qualitatively, the graph of $\partial F / \partial (V/V_R)$ sketched in Fig. 5.2 is like that found in such experiments. As is discussed by Beatty [23], it is a good idea to break in a balloon before performing measurements. Blow it up and deflate it once or twice. If one does not, the graph is likely to be quite a bit higher on the first run than on subsequent runs. After break in, the experimental curves come closer to being reproducible and reversible, features one would like to have. In such respects, cheap toy balloons will not perform very well as a rule and a careful experimentalist will want better materials. However, the general phenomena indicated are quite robust, occurring in balloons from various sources which may or may not be very close to spherical. It is related to a common experience. As one starts blowing up a balloon, at first one must blow quite hard to increase the size a little. Then, rather suddenly, it becomes easier as one goes over the hill in the p–V diagram.

Then, it gets harder again until the balloon decides to burst. It is hardly reasonable to use the same analysis for long tubular balloons, although one observes somewhat similar phenomena, along with aneurisms which develop and grow in length as air is added.

To simplify analyses, we assume that the balloons stay spherical. Instabilities could occur, inducing the balloon to take on some less symmetrical shape. Later, we will discuss some evidence suggesting that the possibility is real. However, it is a more difficult problem to analyze this, and, for the experimentalist, it is hard to get balloons that are very good spheres of uniform thickness, and so on. It is a matter of experience that, near conditions at which instabilities occur, small differences of this kind can have a significant effect. Such quirks are discussed by Thompson and Hunt [25].

To get an idea of the number and character of equilibria, it helps to consider another picture. First consider the graph of $\partial F/\partial V$, regarded as a function of V. This differs from that given in Fig. 5.2 only by a simple change of scale. For simplicity, assume that we use the ideal gas model, described by (5.1.11) with $V_G = V$. With V as abscissa and p as ordinate add in the p–V curves, in this case hyperbolae, for various values of M. From (5.2.4), the possible equilibria are given by the points where one of these intersect the graph for the balloon. From such a sketch, one can see that there is at least a theoretical possibility that by adjusting M, one might get one, two or three equilibria.[2] Consider the picture in Fig. 5.3 when there are three, giving $V_1 < V_2 < V_3$ as equilibrium volumes.

Now, as we did before in considering the equal area rule, we can picture energy differences as area differences. Obviously

$$
A_1 = F\left(\frac{V_2}{V_R}\right) - F\left(\frac{V_1}{V_R}\right) - \int_{V_1}^{V_2} p\,dV
$$

$$
-A_2 = F\left(\frac{V_3}{V_R}\right) - F\left(\frac{V_2}{V_R}\right) - \int_{V_2}^{V_3} p\,dV
$$

(5.2.6)

with the latter integrals referring to the gas. However, with (5.1.9) we can replace these by differences in F_G, for example

$$
-\int_{V_1}^{V_2} p\,dV = F_G\left(\frac{V_2}{M}\right) - F_G\left(\frac{V_1}{M}\right).
$$

(5.2.7)

---

[2]By adjusting the shape of the rubber curve, as indicated in Fig. 5.3, you can get a rather similar sketch where the two curves never intersect more than once. This is a way of constructing cases where (5.2.5) always holds. According to Alexander [26], this type of behavior is encountered in some neoprene balloons used for high-altitude measurements.

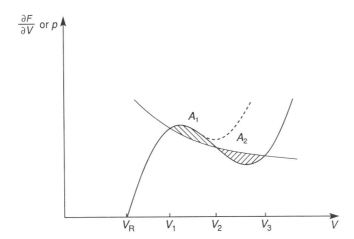

FIGURE 5.3. Sketch of the balloon and gas response curves, for a value of M giving three points of intersection. Here A_1 and A_2 denote the areas of the cross-hatched regions. The dashed curve indicates another possible graph for the balloon giving only one point of intersection with the graphs for the gas.

Then, with (5.2.2) we have

$$
\begin{aligned}
E(V_2, M) - E(V_1, M) &= A_1 > 0 \\
E(V_2, M) - E(V_3, M) &= A_2 > 0.
\end{aligned}
\qquad (5.2.8)
$$

Thus, the equilibrium at $V = V_2$ is less stable than those at V_1 or V_3. It is in fact unstable, as can be seen by doing similar area comparisons, for V near V_2. The stability enhancement indicated by (5.2.5) is associated with the possibility of V_1 taking values where $\partial^2 F/\partial V^2 < 0$, as it can. Again, area comparisons indicate that, as long as we have the three equilibria, V_1 and V_3 are both at least relative minima. From (5.2.8) one can read off which is stable:

$$
\begin{aligned}
A_1 \geq A_2 &\Rightarrow V_1 \text{ stable}, \\
A_1 \leq A_2 &\Rightarrow V_3 \text{ stable}.
\end{aligned}
\qquad (5.2.9)
$$

So, we do have an "equal area" rule somewhat similar to those discussed earlier, although the areas are not defined by intersecting the basic graph with a straight line ($p = $ const.). With a single spherical balloon, we do not have a physically reasonable analogue of the phase mixtures encountered in bars and plates. One begins to obtain something similar if one considers several balloons, interconnected so gas is free to move from one to another.

Here, since pV is proportional to M for an ideal gas, increasing M moves the p–V curve outward, in an obvious sense. If M is small enough, it will intersect the graph just once, to the left. As M increases, it will attain a value where the p–V curve touches the graph: here V_2 and V_3 just appear, with $V_2 = V_3$. As M increases further, these split, V_2 becoming unstable, V_3 metastable. Further increasing M brings us to the equal area configuration, where V_1 gives way to V_3, becoming the metastable configuration. Increasing M then moves V_1 and V_2 together: V_1 stays metastable until they coincide. For greater values of M there is again just one possible equilibrium configuration. Clearly, the stability picture here is quite different from what it is when pressure is controlled, although it is not hard to see that we will be left with some values of V too unstable to be observed, when Fig. 5.3 applies.

Of course, this picture also applies to possible equilibria with $h = 0$ in the balloon problem first considered. It is somewhat complicated by the need to account for (5.1.22), a limitation on pressure which depends on the weight, along with the possibility that the most stable configurations may occur when $h > 0$. We will not pursue analysis of this: the ideas needed to do so have been covered.

5.3   Exercises

5.1. To check the mathematical result used in (5.1.20), consider

$$f(x, y) = ax^2 + 2bxy + cy^2,$$

where a, b, and c are constants. Show that, to have $f(x, y) \geq 0$ for all x and y, it is necessary and sufficient that

$$a \geq 0, \quad c \geq 0, \quad b^2 \leq ac.$$

5.2. Show that (5.1.18) and (5.1.20) give the same results when $h > 0$. Do they when $h = 0$?

5.3. For a spherical balloon at fixed temperature, the Neo-Hookean theory of rubber yields a total strain energy function of the form

$$F = a(2\lambda^2 + \lambda^{-4} - 3), \qquad \lambda = \frac{r}{R},$$

with a a positive constant. Make a sketch of the corresponding graph of the pressure-volume function. For cases where it is subject to constant pressure, calculate the value of V/V_0 representing a limit of metastability and say what you can about the possibility of stable equilibria.

5.4. Another commonly used theory of rubber is the Mooney–Rivlin theory. For a spherical balloon,

$$F = a[2\lambda^2 + \lambda^{-4} - 3 + (2\lambda^{-2} + \lambda^4 - 3)/K],$$

where a and K are positive constants. Fitting data on various kinds of rubber generally gives values of K in the range $4 \leq K \leq 8$. Make a sketch of the graph of the pressure-volume function for a realistic value of K. Is there any important difference between this and the previous graph, and, if so, what is it?

5.5. In terms of three-dimensional theory, the above estimates of F are made using an approximation which is reasonable when

$$\frac{R_2 - R_1}{R_1} << 1.$$

More accurately, for the neo-Hookean theory,

$$F = 4\pi \int_{R_1}^{R_2} W R^2 \, dR,$$

where W, the strain energy per unit volume, is the function

$$W = C(2\lambda^2 + \lambda^{-4} - 3).$$

Here, C is a positive constant, depending only on the type of rubber. For a very thin balloon, relate the constant a in Exercise (5.3) to C and quantities relating to the geometry of the balloon. Here, volume refers to that of the unstressed material.

6

Biaxial Stretch of Rubber Sheets

6.1 The Idealized Problem

We now consider biaxial stretch experiments performed on rubber, one of the isothermal mechanical experiments done to obtain information about the relevant strain energy functions.

Again, we choose as a reference a stable unloaded configuration, say at room temperature. For simplicity, we will only consider experiments at this temperature. In the reference configuration, the sheet occupies a region described by

$$-a \le x_1 \le a, \qquad -a \le x_2 \le a, \qquad -b \le x_3 \le b, \qquad (6.1.1)$$

where (x_1, x_2, x_3) are coordinates in a suitable selected rectangular Cartesian coordinate system. Generally, b/a is quite small; thin sheets are commonly used. The aim is to produce simple kinds of deformations by applying suitable loads to the edges of the sheet. Briefly, the material point at (x_1, x_2, x_3) goes to (y_1, y_2, y_3) with

$$y_1 = \lambda_1 x_1, \qquad y_2 = \lambda_2 x_2, \qquad y_3 = \lambda_3 x_3, \qquad (6.1.2)$$

where the stretches λ_i are positive constants. That is, we will only consider deformations of this kind. Here, the assumption is that the strain energy function W is a function of λ_i, satisfying

$$W(\lambda_1, \lambda_2, \lambda_3) = W(\lambda_2, \lambda_1, \lambda_3) = W(\lambda_1, \lambda_3, \lambda_2) \qquad (6.1.3)$$

and

$$W(1,1,1) = 0, \qquad (6.1.4)$$

this being interpreted as energy per unit reference volume. To within an additive constant this is also the Helmholtz free energy function at the same temperature, the analogue of the relation between such quantities we had in our earlier discussion of thermodynamic experiments on bars, in Section 2.5. That (6.1.3) should hold is inferred from the fact that rubber is an isotropic material. Here, as in our considerations of Martensitic transformations, considerations relating to material symmetry will play an important role.

As noted in our discussion of balloons, it is reasonable to consider rubbers as incompressible materials, which means that the λ_i should always satisfy the equation

$$\lambda_1 \lambda_2 \lambda_3 = 1. \qquad (6.1.5)$$

Using this, we can replace W by a function of two variables, setting

$$U = U(\lambda_1, \lambda_2) = W\left(\lambda_1, \lambda_2, \frac{1}{\lambda_1 \lambda_2}\right). \qquad (6.1.6)$$

Then (6.1.3) implies that

$$U(\lambda_1, \lambda_2) = U(\lambda_2, \lambda_1) = U\left(\frac{1}{\lambda_1 \lambda_2}, \lambda_2\right). \qquad (6.1.7)$$

Ideally, we would like to apply forces on the edges which are normal to the edges and uniformly distributed, with equal and opposite forces on opposite edges, leaving the faces $x_3 = \pm b$ free. On the faces $x_1 = \pm a$, we can then represent the resultant forces by

$$F_1^+ = 4T_1 ab, \qquad (6.1.8)$$

and

$$F_1^- = -4T_1 ab, \qquad (6.1.9)$$

where T_1 is a number representing force per unit reference area, a stress. The forces on $x_2 = \pm a$ are similarly represented by a number T_2.

We consider an ideal dead-load device, able to control these forces to keep T_1 and T_2 at fixed values. Consider changing y_i to $y_i + \Delta y_i$. On the face $x_1 = a$, this will produce the displacement

$$\Delta y_1 = (\Delta \lambda_1) a,$$

so the work done here by F_1^+ is

$$F_1^+ \Delta y_1 = 4a^2 b T_1 \Delta \lambda_1.$$

Adding the similar contributions from the other three faces, we get as the total

$$8a^2 b(T_1 \Delta \lambda_1 + T_2 \Delta \lambda_2) = 8a^2 b \Delta(T_1 \lambda_1 + T_2 \lambda_2),$$

since T_1 and T_2 are here held fixed. This means that the loading device can be considered to be conservative with the potential energy

$$\chi = -8a^2b(T_1\lambda_1 + T_2\lambda_2).$$

Of course, we have in mind that this system is in contact with a heat bath at room temperature, so we have a situation fitting statement IV in Chapter 1. However, we will take the usual shortcut, assuming that the sheet, etc., is at room temperature, using the Helmholtz free energy in place of the ballistic free energy. For the sheet itself, we can use the total strain energy

$$U(2a)(2a)(2b) = 8a^2bU.$$

After cancelling the numerical factor $8a^2b$, this gives the relevant thermo-dynamic potential as

$$E = U(\lambda_1, \lambda_2) - T_1\lambda_1 - T_2\lambda_2. \tag{6.1.10}$$

As usual, stable or metastable equilibria correspond to absolute or relative minima of this function of λ_1 and λ_2 for fixed values of T_1 and T_2 and, here, physically possible variations include arbitrary small changes in these variables. Again, there is a possibility that we will miss some instabilities because we only allow for some very simple kinds of deformation. Physically, some kinds of loads can easily make a thin sheet undergo more complicated buckling deformations beyond the scope of our theory.

With (6.1.10), the first derivative test gives the equilibrium equations

$$T_1 = \frac{\partial U}{\partial \lambda_1}, \qquad T_2 = \frac{\partial U}{\partial \lambda_2}, \tag{6.1.11}$$

the idea being to try to solve this for λ_1 and λ_2 for given values of T_1 and T_2 and to discard those corresponding to unstable equilibria. As before, we are concerned with a function of two variables and the second derivative test gives us the conditions analogous to (5.1.20), which are

$$\frac{\partial^2 U}{\partial \lambda_1^2} \geq 0,$$

$$\frac{\partial^2 U}{\partial \lambda_2^2} > 0, \tag{6.1.12}$$

$$\left(\frac{\partial^2 U}{\partial \lambda_1 \partial \lambda_2}\right)^2 \leq \frac{\partial^2 U}{\partial \lambda_1^2}\frac{\partial^2 U}{\partial \lambda_2^2}.$$

For stable equilibrium, we find values $\bar{\lambda}_1$ and $\bar{\lambda}_2$ satisfying (6.1.11) such that, for any possible value of λ_1 and λ_2,

$$U(\lambda_1, \lambda_2) - T_1\lambda_1 - T_2\lambda_2 \geq U(\bar{\lambda}_1, \bar{\lambda}_2) - T_1\bar{\lambda}_1 - T_2\bar{\lambda}_2. \tag{6.1.13}$$

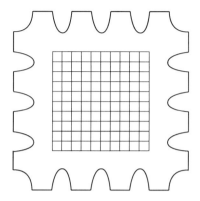

FIGURE 6.1. Sketch of specimen used in biaxial stretch experiments.

Associated with this is a geometrical picture similar to that associated with the one-dimensional theory of bars, etc., under dead loads. Consider $(\lambda_1, \lambda_2, U)$ as rectangular Cartesian coordinates in three-dimensional space. Then $U = U(\lambda_1, \lambda_2)$ is the equation of a surface. On this is the point $(\overline{\lambda}_1, \overline{\lambda}_2, \overline{U} = U(\overline{\lambda}_1, \overline{\lambda}_2))$. Here, the tangent plane to the surface has the equation

$$U - \overline{U} - T_1(\lambda_1 - \overline{\lambda}_1) - T_2(\lambda_2 - \overline{\lambda}_2) = 0, \qquad (6.1.14)$$

when $\overline{\lambda}_1$ and $\overline{\lambda}_2$ satisfy (6.1.11). Points lying above this plane have coordinates satisfying the inequality

$$U - \overline{U} - T_1(\lambda_1 - \overline{\lambda}_1) - T_2(\lambda_2 - \overline{\lambda}_2) > 0,$$

and those lying below it have coordinates satisfying the reverse inequality. Thus, (6.1.13) means that every point on the surface must lie in the half-space above the tangent plane. For $U(\lambda_1, \lambda_2)$ to be a convex function means that it does lie in this half-space for the tangent planes at all points on the surface.

With the same assumptions concerning the allowed deformations, the theory of hard devices is trivial. That is, if we fix the values of y_i on the edges, we will have either an impossible situation or only one possible deformation. For example, on $x_1 = a$, $y_1 = \lambda_1 a$, so fixing y_1 fixes λ_1, etc. So, we are left with no possible variations, implying that any possible configuration is stable.

Although our formulation is very simplistic, it is good enough to provide a basis for beginning to understand some interesting phenomena occurring in rubber.

Experimentally, one cannot come very close to matching the ideal conditions assumed above. Earlier devices were designed to approximate dead loading. The test specimens had fillets on the edges, a square grid marked on the interior, as indicated in Fig. 6.1.

To the lugs are attached fairly long strings or wires, the other ends being connected to a square frame in such a way that one can adjust them to vary the force exerted by them on the specimen. One makes adjustments to get the grid lines straight and perpendicular as they should be for the kinds of deformations we have assumed. This is a tedious matter but, with some experience, one can get what appears to be the correct kind of deformation in a sizeable part of the sheet and hence estimate values of λ_1 and λ_2. From measurements of the forces supplied through the strings, one then estimates the resultant forces applied to edges of this part, obtaining an estimate of T_1 and T_2. These are always tensile forces ($T_i \geq 0$). At least to some degree, this helps to avoid the buckling which is more associated with compressive loads. Various modifications in design were tried, the trend being to make the devices harder to come closer to controlling the edge displacements. Our crude theory suggests that this may help to extend the range of values of λ_1 and λ_2 which can be observed by suppressing instabilities which may occur in softer devices, producing deformations of a more complicated kind. I do not know of clear experimental evidence supporting this view. Possibly related to this is a curious matter discussed in the next section.

6.2 The Treloar Instability

In 1948, Treloar [27] published some data on rubbers obtained using a biaxial stretch device of the old kind.[1] Although he did not say much about it, some of the measurements were, in one respect, curious. Intuitively, one expects that when $T_1 = T_2$, we will have $\lambda_1 = \lambda_2$. In some cases, his data for situations where $T_1 = T_2$ indicated that $\lambda_1 \neq \lambda_2$ and the difference seems too large to be attributed to the inevitable experimental errors. This appears to have been overlooked or forgotten by rubber workers until 1986, when Kearsley [28] discussed it and some theoretical reasons for believing that this phenomenon is real. Let us consider a slightly different analysis of it, indicating that it is, in some respects, similar to the Martensitic transformations discussed before.

For this purpose, it is convenient to make a change of variables. Set

$$\begin{matrix} \lambda_1 = \delta + \gamma \\ \lambda_1 = \delta \quad \gamma \end{matrix} \Rightarrow \begin{matrix} \delta = (\lambda_1 + \lambda_2)/2 \\ \gamma = (\lambda_1 \quad \lambda_1)/2 \end{matrix} \Rightarrow \delta > |\gamma|, \qquad (6.2.1)$$

and set

$$U(\lambda_1, \lambda_2) = V(\delta, \gamma). \qquad (6.2.2)$$

Then, from (6.1.7) we have, in particular,

$$V(\delta, \gamma) = V(\delta, -\gamma). \qquad (6.2.3)$$

---

[1] Included in Treloar's paper is a photograph of a loaded sample.

You may explore for yourself the other implications. Also, (6.1.10) now becomes

$$E = V - \pi\delta - \tau\gamma, \tag{6.2.4}$$

where

$$\pi = T_1 + T_2, \qquad \tau = T_1 - T_2, \tag{6.2.5}$$

and the equilibrium equations now become

$$\pi = \frac{\partial V}{\partial \delta}, \qquad \tau = \frac{\partial V}{\partial \gamma}. \tag{6.2.6}$$

With V an even function of γ and τ related to it as the derivative with respect to γ, τ and γ become analogous to the shear stress and shear strain considered in the theory of Martensitic transformations. That another variable δ is involved does complicate matters somewhat. However, its variations may be considered to be similar to the variations in θ considered before.

In these terms, Treloar's data indicate that, for some values of δ, $\tau = 0$ but $\gamma \neq 0$, giving us an analogue of Martensite. It is not likely that this is true for all δ. For example, indications are that there is only one unloaded state, so $\pi = \tau = 0 \Rightarrow \delta = 1$ and $\gamma = 0$, giving us an analogue of Austenite. Indications are that something more or less like a Martensitic transformation should be encountered in some loading programmes. One might expect to see something similar to twinning but experimentalists seem not to have reported this. Theoretically, one can understand this. If one tries to construct deformations involving jump discontinuities in γ, even allowing them in δ, one finds that it is impossible to have all three components of displacement continuous. Thus, in this respect, the analogy with the theory of Martensitic transformations breaks down. Conversations with some rubber experts indicated they were unaware of the possibility of such an instability, until Kearsley revived the issue and his work is not yet widely known.

Occasionally, something small can produce a surprising error in an experiment, so it is important to try to make an assessment of this situation theoretically. One approach is to take a form of W which can be analyzed and fits some relevant data, at least roughly. For rough calculations, workers often use the Mooney–Rivlin form,

$$W = C_1(\lambda_1^2 + \lambda_2^2 + \lambda_3^2 - 3) + C_2\left(\frac{1}{\lambda_1^2} + \frac{1}{\lambda_2^2} + \frac{1}{\lambda_3^2} - 3\right),$$

with C_1 and C_2 constants, determined by curve fitting. What is important for our purposes is their ratio, so we will use the simpler form

$$W = K(\lambda_1^2 + \lambda_2^2 + \lambda_3^2 - 3) + \frac{1}{\lambda_1^2} + \frac{1}{\lambda_2^2} + \frac{1}{\lambda_3^2} - 3. \tag{6.2.7}$$

This gives

$$V = K \left[2(\delta^2 + \gamma^2) + \frac{1}{(\delta^2 - \gamma^2)^2} - 3 \right]$$

$$+ \frac{1}{(\delta + \gamma)^2} + \frac{1}{(\delta - \gamma)^2} + (\delta^2 - \gamma^2)^2 - 3.$$

(6.2.8)

Then a calculation gives

$$\tau = \frac{\partial V}{\partial \gamma} = \frac{4\gamma}{(\delta^2 - \gamma^2)^3} = \left\{ K[(\delta^2 - \gamma^2)^3 + 1] + \gamma^2 + 3\delta^2 - (\delta^2 - \gamma^2)^4 \right\}. \quad (6.2.9)$$

One of the conditions for metastability is that $\partial^2 V / \partial \gamma^2 \geq 0$. Let us explore this at $\gamma = 0$, using

$$\left(\frac{\partial^2 V}{\partial \gamma^2} \right)_{\gamma = 0} = \lim_{\gamma \to 0} \frac{1}{\gamma} \frac{\partial V}{\partial \gamma}$$

$$= \frac{4}{\delta^6} [(K(\delta^6 + 1) + 3\delta^2 - \delta^8]$$

(6.2.10)

$$= \frac{4(\delta^6 + 1)}{\delta^6} [K - f(\delta)],$$

where

$$f(\delta) = \frac{\delta^2 (\delta^6 - 3)}{\delta^6 + 1}, \quad (6.2.11)$$

is rather easy to visualize, from its graph.

For a positive value of K, such as that indicated in Fig. 6.2, there is just one value of $\delta = \bar{\delta}$ for which $f = K$, and we have

$$K > f(\delta) \text{ for } \delta < \bar{\delta}, \qquad K < f(\delta) \text{ for } \delta > \bar{\delta},$$

so we have instability at $\gamma = 0$, when $\delta > \bar{\delta}$, and when $\delta < \bar{\delta}$ the one stability inequality is satisfied. Curve fitting produces various values of K depending on the particular rubber and the range of deformation which one selects to fit best. Kearsley mentions that likely values range from 4 to 9, although one can find estimates outside this range. Roughly, this puts $\bar{\delta}$ somewhere between 2 and 3, not unreasonably large for rubber. Also, this means that for any fixed $\delta < \bar{\delta}$ and γ close enough to zero, V is a convex function of γ and this is not true for $\delta > \bar{\delta}$. Thus, for fixed $\delta < \bar{\delta}$, τ is a monotonically increasing function of γ, suggesting that we have an analogy with the second-order Martensitic transformations. To consider other possible solutions of $\tau = 0$, we note that they will occur when

$$K[(\delta^2 - \gamma^2)^3 + 1] + \gamma^2 + 3\delta^2 - (\delta^2 - \gamma^2)^4 = 0.$$

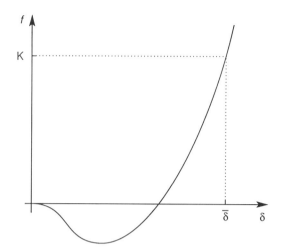

FIGURE 6.2. Graph of the function given in equation (6.2.11).

Set

$$\hat{\delta} = \sqrt{\delta^2 - \gamma^2} \le \delta,$$

and the last equation can be rewritten in the form

$$\frac{4\gamma^2}{\hat{\delta}^6 + 1} + K - f(\hat{\delta}) = 0. \tag{6.2.12}$$

Bearing in mind that $\delta^2 > \gamma^2$, the first term is positive, so the sum of the other two terms must be negative. From above, this means that we must have

$$\delta \ge \hat{\delta} \ge \overline{\delta}.$$

Thus, for any fixed $\delta < \overline{\delta}$, $\tau = 0$ implies $\gamma = 0$. Consider a fixed $\delta > \overline{\delta}$. It is easily checked that the first term in (6.2.12) is an increasing function of γ^2. As γ^2 increases, $\hat{\delta}$ decreases, as does $f(\hat{\delta})$. Thus the entire expression increases with γ^2. At $\gamma = 0$, it is negative and it is positive when $\sqrt{\delta^2 - \gamma^2} = \overline{\delta}$. So, it must vanish for one value of γ^2 between these values and for no other values of γ^2. By a similar argument, the values of $\hat{\delta}$ satisfying (6.2.12) must increase as γ^2 increases. This also makes the behaviour with δ increasing rather like that which we saw with decreasing θ in the Martensitic transformations, when the latter are of second-order. For $\delta > \overline{\delta}$, drawing a graph of τ versus γ with a negative slope at $\gamma = 0$, with $\tau = 0$ at one other value of γ^2 gives a graph at least roughly like that suggested by the analogy, as shown in Fig. 6.3.

By a rather similar argument, one can show that for fixed $\delta < \overline{\delta}$, τ is a monotonically increasing function of γ. Of course, it is possible that, for a more realistic form of the strain energy function, one may get a picture more like that suggested by the first-order Martensitic transformations or something else. Other arguments given by Kearsley make it seem very unlikely

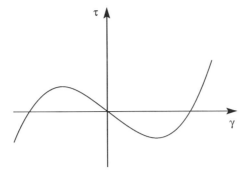

FIGURE 6.3. Graph of τ vs γ, for fixed $\delta > \bar{\delta}$.

that one will not have some such instability. It is curious that experimentalists working more recently than Treloar have not noticed, or at least not reported, something unusual associated with this instability. Possibly, by employing harder devices, they have eliminated such instabilities. Possibly, some would have spotted something if they knew what to look for. It is also possible that some dealt with limited ranges of stretch which happened not to include these instabilities.

For the model we are exploring we have not completed checking the inequalities governing stability or metastability. We do know that the nontrivial solutions of $\tau = 0$ give values of γ^2 which decrease as we decrease δ until they become zero at $\delta = \bar{\delta}$. This means that the plot of the curves $\tau = 0$ in the δ–γ plane then looks much like the pitchfork discussed in Section 4.2.

For stability, it is also important that we have

$$\frac{\partial^2 V}{\partial \delta^2} = \frac{\partial \pi}{\partial \delta} \geq 0. \tag{6.2.13}$$

A simple calculation shows that the strict inequality holds for all possible values of δ and γ. This means that, for fixed γ, V is a convex function of δ, with π a monotonically increasing function of δ. Thus we can, in principle, solve

$$\pi = \frac{\partial V}{\partial \delta},$$

giving

$$\delta = g(\pi, \gamma), \tag{6.2.14}$$

although it is not easy to do this explicitly.

To check the remaining inequality

$$\frac{\partial^2 V}{\partial \gamma^2} \frac{\partial^2 V}{\partial \delta^2} \geq \left(\frac{\partial^2 V}{\partial \delta \partial \gamma} \right)^2 \tag{6.2.15}$$

is not so easy. It is simple to check this when $\gamma = 0$ since the right side vanishes. Thus (6.2.15) holds, strictly when $\delta < \bar{\delta}$, and it fails when $\delta > \bar{\delta}$. By continuity, it will also fail for $\delta > \bar{\delta}$, when γ is nonzero, but sufficiently small.

Now, a straightforward calculation gives, for $\gamma \neq 0$,

$$\frac{\pi}{\delta} - \frac{\tau}{\gamma} = 2\left(\delta^2 - \gamma^2 - \frac{K}{\delta^2 - \gamma^2}\right), \tag{6.2.16}$$

which increases with $(\delta^2 - \gamma^2)$. Consider the outer tine of the pitchfork with $\gamma > 0$. On it, as γ increases so does γ^2 and, as already noted after (6.2.12), so also will δ and $(\delta^2 - \gamma^2)$. On it, δ will be a function of γ, with

$$\frac{d\tau}{d\gamma} = \frac{\partial\tau}{\partial\gamma} + \frac{\partial\tau}{\partial\delta}\frac{d\delta}{d\gamma} = 0. \tag{6.2.17}$$

From Fig. 6.3, it is clear that here

$$\frac{\partial\tau}{\partial\gamma} = \frac{\partial^2 V}{\partial\gamma^2} > 0, \tag{6.2.18}$$

one of the inequalities needed for stability. Another is $\partial^2 V/\partial\delta^2 > 0$, which, we have noted, is always satisfied. Since δ increases with γ, (6.2.17) can only be satisfied if

$$\frac{\partial\tau}{\partial\delta} = \frac{\partial^2 V}{\partial\gamma\partial\delta} = \frac{\partial\pi}{\partial\gamma} < 0. \tag{6.2.19}$$

We also know that, on this curve,

$$\frac{d\pi}{d\gamma} = \frac{\partial\pi}{\partial\gamma} + \frac{\partial\pi}{\partial\delta}\frac{\partial\delta}{\partial\gamma}$$

$$= \frac{\partial^2 V}{\partial\gamma\partial\delta} + \frac{\partial^2 V}{\partial\delta^2}\frac{d\delta}{d\gamma} > 0. \tag{6.2.20}$$

Look at (6.2.16), bearing in mind that δ and $(\delta^2 - \gamma^2)$ increase with γ on this curve. Solving for $d\delta/d\gamma$, and substituting this in (6.2.20) gives

$$\frac{\partial^2 V}{\partial\gamma\partial\delta} - \frac{\partial^2 V}{\partial\delta^2}\frac{\partial^2 V}{\partial\gamma^2} \Big/ \frac{\partial^2 V}{\partial\gamma\partial\delta} > 0.$$

Multiplying this by the negative quantity $\partial^2 V/\partial\gamma\partial\delta$ then gives the third stability inequality listed in (6.1.12),

$$\left(\frac{\partial^2 V}{\partial\gamma\partial\delta}\right)^2 - \frac{\partial^2 V}{\partial\delta^2}\frac{\partial^2 V}{\partial\gamma^2} < 0. \tag{6.2.21}$$

So, on this branch, we have equilibria which are at least metastable. Actually, they are stable. A similar argument gives the same conclusion for the other outer tine, on which $\gamma < 0$. In between them, somewhere near $\gamma = 0$, is a region of instability. Qualitatively, the picture is then clear, and we cannot trust the constitutive equation to give quantitatively correct estimates of the situation.

Assuming the estimate is qualitatively correct, it suggests a strategy for designing experiments to learn more about the instabilities. Roughly, we would like to devise loading programs to aim at the boundary of the unstable region and find out what really happens. It may be profitable to use a soft device, one of the older designs being better than the newer. If we start with the unloaded configuration, keep $\tau = 0$ and increase π, δ should increase with γ staying zero for a while; but, if δ becomes large enough, we should hit the instability at $\delta = \bar{\delta}, \gamma = 0$. The simplest possibility is that we continue to have the simple kind of deformation assumed, but move along the curve path $\tau = 0$, γ becoming nonzero. Treloar seems to have hit points on $\tau = 0, \gamma \neq 0$, accidentally, with deformations fitting our assumptions, so this could well happen. "Pitchfork bifurcations" more or less like this are encountered in various physical systems. Now, by making $\tau \neq 0$, we can try to head back towards $\gamma = 0$ and should encounter other points on the boundary of the unstable region and see something unusual happen. Here, by thinking hard about a somewhat shaky theory, we have been led to the design of an experiment which is likely to produce useful information.

There is a loosely related matter. In considering the balloon problems, we assumed that they remain spherical. Intuitively, this is associated with the notion that at any point on the balloon, the material should be stretched equally in all tangent directions, the amount of stretch being the same at all points. Roughly, this is similar to assuming homogeneous deformation in the sheets, with $\gamma = 0$ when $\tau = 0$. We now have some reason to question such assumptions, although the conditions for stability are somewhat different in the two cases. If one blows up toy balloons of a symmetrical shape, more or less, sphere-like, one does see the development of curious asymmetries. One can argue that such effects stem from variations in thickness, etc. However, one rubber expert, who had made a strong effort to obtain the best possible spherical balloons, told me that he is convinced that this is a real effect. Also, Alexander [26] presents a combination of theoretical and experimental evidence that the effect is real in some neoprene balloons. The analysis of Haughton and Ogden [29] suggests that it might not be true for thicker balloons in the case where the pressure is controlled. Bear in mind that rather thick balloons are what we often call balls and ask yourself what your experience suggests about the difference between balls and balloons.

The behavior of some biological materials is rather like rubber. Estimates of strain energy functions I have seen for some lung tissue suggest that a similar instability may occur in these, although this does involve extrapolating the function to larger stretches than were observed.

For the experiment at hand, allowing for more complicated deformations does change the conditions required for stability to some degree. The best available analyses of this are by Chen [30, 31]; the latter deals with hard devices and does indicate that the stability conditions for these are different.

One of the best general references on the physics of rubber is the book by Treloar [32].

6.3 Exercises

6.1. Determine whether a strain energy function of the Neo-Hookean form

$$W = C(\lambda_1^2 + \lambda_2^2 + \lambda_3^2 - 3), \qquad C > 0$$

is capable of describing the Treloar instability.

6.2. In practice, our sheet might be called a bar, if two of its dimensions are small compared to the third; replace (6.1) by

$$-a \le x_1 \le a, \qquad -b \le x_2 \le b, \qquad -c \le x_3 \le c,$$

with

$$b/a << 1, \qquad c/a << 1.$$

With this, bar theory would identify the stretch λ as λ_1 and assume that forces are applied only in the x_1 direction, so take $T_2 = 0$. For the Mooney–Rivlin form of W, determine the corresponding form of the strain energy function for bar theory.

6.3. For the bar theory just deduced, is the strain energy function a convex function for positive values of λ and realistic values of K? From your experience with bars, what would infer about possible instabilities that might be predicted by such theory?

6.4. In Section 6.1, it was asserted that points lie above or below the tangent plane given by (5.1.14), depending on whether

$$U - \bar{U} - T_1(\lambda_1 - \bar{\lambda}_1) - T_2(\lambda_2 - \bar{\lambda}_2)$$

is positive or negative. Prove or disprove this.

6.5. In Section 6.1, after (6.2.12), it was asserted that "the value of $\hat{\delta}$ satisfying (6.2.12) must increase as γ^2 increases." Prove or disprove this.

6.6. Take a toy balloon of spherical shape. What do you think would happen to a great circle drawn on it, when it is inflated just enough to do this with a marking pen, then inflated more? Try this experiment, and you might see why some workers have concluded that the actual deformation is more complicated than was assumed in our balloon studies, even when the balloon appears to stay close to being a sphere. Developing better theory is not an elementary exercise, as is often the case in stability studies.

7
Moving Discontinuities

7.1 Shock Waves in Bars

We have encountered examples of static discontinuities, for example the twins in Martensite or the "phase mixtures" in bars. Physically, such discontinuities are often better regarded as thin regions through which the quantities of interest vary smoothly, but rapidly. However, simpler kinds of equations sometimes permit us to model them better if we treat them as discontinuities. Such phenomena are likely to occur dynamically, when one suddenly imposes a load on a solid, by an impact, or sometimes, by setting off an explosive on its surface. Often, experiments of this kind along with one-dimensional theory are used to provide information concerning constitutive equations for solids. It is common to use thermodynamic considerations. Here, we will discuss commonly used ideas, in the context of bar theory.

In this context a "surface" of discontinuity becomes a point

$$x \to \psi(t), \tag{7.1.1}$$

moving through the material with velocity

$$v = \dot{\psi}(t), \tag{7.1.2}$$

which we assume is positive. This is called the material or referential wave speed, or, more briefly, the wave speed. Some functions of interest are assumed to have jump discontinuities at this point. If $g = g(x, t)$ is such a function, we can fix t and consider the limit as we approach the point from the front, the region into which the wave is moving, to define

$$g^+(t) = \lim_{x \to \psi^+(t)} g(x, t). \tag{7.1.3}$$

Similarly, we have the limit from the opposite side and the abbreviated notation for the jump indicated by

$$g^-(t) = \lim_{x \to \psi^-(t)} g(x, t), \qquad [g] = g^+ - g^-. \tag{7.1.4}$$

As in the static case, we assume that $y(x, t)$ is continuous, excluding possibilities of breaking, etc., so

$$[y] = 0. \tag{7.1.5}$$

Consider the possibility that the stretch λ and velocity $\dot{y}$ have finite discontinuities. Evaluating $y(x, t)$ on the front and back sides and differentiating with respect to t, we then obtain

$$\begin{aligned} dy^+/dt &= \lambda^+ v + \dot{y}^+ \\ dy^-/dt &= \lambda^- v + \dot{y}^-. \end{aligned} \tag{7.1.6}$$

Differentiating (7.1.5) with respect to t then gives

$$d[y]/dt = [\lambda]v + [\dot{y}] = 0 \tag{7.1.7}$$

as a kinematic condition, restricting these jumps.

With the discontinuities, it is not sufficient to satisfy the differential equations of motion, etc., where things are smooth enough to do so, although this should be done. Consider the integral forms. One is the usual mechanical equation

$$\frac{d}{dt} \int_{x_1}^{x_2} \rho \dot{y} \, dx = \int_{x_1}^{x_2} f \, dx + \sigma \Big|_{x_1}^{x_2} \tag{7.1.8}$$

with x_1 and x_2 chosen so that, at a particular time, $x_1 < \psi(t) < x_2$. Recall that ρ is a constant. We assume that, if x_1 and x_2 are close enough to $\psi(t)$, the functions are smooth, except at $x = \psi(t)$. Then, on the left, we have

$$\frac{d}{dt} \left(\int_{x_1}^{\psi} \rho \dot{y} \, dx + \int_{\psi}^{x_2} \rho \dot{y} \, dx \right) = \int_{x_1}^{x_2} \rho \ddot{y} \, dx + \rho \dot{y}^- v - \rho \dot{y}^+ v \tag{7.1.9}$$

or

$$\frac{d}{dt} \int_{x_1}^{x_2} \rho \dot{y} \, dx = \int_{x_1}^{x_2} \rho \ddot{y} \, dx - \rho v [\dot{y}]. \tag{7.1.10}$$

Putting this back in (7.1.8) and taking the limit as $x_1 \to \psi^-$, $x_2 \to \psi^+$, we obtain

$$-\rho v [\dot{y}] = [\sigma], \tag{7.1.11}$$

as another restriction on jumps. By a similar analysis, the energy equation

$$\frac{d}{dt} \int_{x_1}^{x_2} \left(\varepsilon + \frac{1}{2}\rho\dot{y}^2 \right) dx = \int_{x_1}^{x_2} (f\dot{y} + r) \, dx + (\sigma\dot{y} + q) \Big|_{x_1}^{x_2} \tag{7.1.12}$$

gives rise to

$$-v \left[\varepsilon + \frac{1}{2}\rho\dot{y}^2 \right] = [\sigma\dot{y} + q]. \tag{7.1.13}$$

These conditions are, in principle, applicable to plastic wave propagation although in practice, many such waves are not well-idealized as sharp discontinuities. Finally, in a similar way, the Clausius–Duhem inequality can be applied to cases where entropy is well-defined, which does exclude plasticity:

$$\frac{d}{dt} \int_{x_1}^{x_2} \eta \, dx \geq \int_{x_1}^{x_2} \frac{r}{\theta} \, dx + \frac{q}{\theta} \Big|_{x_1}^{x_2}. \tag{7.1.14}$$

It yields the inequality

$$-v[\eta] \geq \left[\frac{q}{\theta} \right]. \tag{7.1.15}$$

From hereon, we assume that our thermoelastic theory of bars applies.[1] Commonly, this is used as one criterion for deciding whether mathematically possible singular solutions are physically acceptable. As is easily checked, if all the equations are satisfied and if we interchange the values (λ^+, η^+) and (λ^-, η^-), the equations are again satisfied with the same v. If the first choice satisfies (7.1.15), the second will not, unless the equality holds in (7.1.15).

Now, using (7.1.7), we can reduce (7.1.11) to the form

$$\rho v^2 = \frac{[\sigma]}{[\lambda]}, \tag{7.1.16}$$

from which it is clear that, for a given material, the wave speed depends on the values of λ and η (or θ) on the two sides. Also, $[\sigma]$ and $[\lambda]$ cannot have opposite signs.

Now, using (7.1.7) and (7.1.11), we have

$$[\sigma\dot{y}] = \sigma^+\dot{y}^+ - \sigma^-\dot{y}^-$$
$$= \frac{\sigma^+ + \sigma^-}{2}(\dot{y}^+ - \dot{y}^-) + \frac{\sigma^+ + \sigma^-}{2}(\dot{y}^+ + \dot{y}^-)$$
$$= \frac{\sigma^+ + \sigma^-}{2}[\dot{y}] - \frac{\sigma^-}{2}(\dot{y}^+ - \dot{y}^-)(\dot{y}^+ + \dot{y}^-)$$
$$= -\frac{\sigma^+ + \sigma^-}{2}v[\lambda] - \frac{v\rho[\dot{y}^2]}{2}.$$

---

[1] Formally, this is analogous to the older one-dimensional theory of shock waves in gases, both being discussed in more detail by Courant and Friedrichs [33]. An exposition by Dunn and Fosdick [34] clears up some old misconceptions about thermoelastic theory of shock waves.

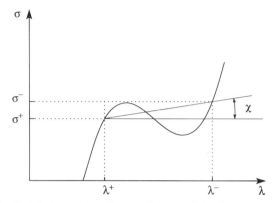

FIGURE 7.1. Sketch indicating the possibility of slowly moving shock waves, in bars with nonmonotone stress–stretch curves.

With this, (7.1.13) reduces to

$$[\varepsilon] - \frac{\sigma^+ + \sigma^-}{2}[\lambda] = -\frac{1}{v}[q]. \tag{7.1.17}$$

Very slowly moving waves can occur as interfaces between different phases or twin planes. As we have seen from static studies, they can even come to rest. In considering cases of this kind, one finds some workers[2] assuming that

$$[\theta] = 0. \tag{7.1.18}$$

Then, it is convenient to introduce the Helmholtz free energy at this common temperature, and put $\varepsilon = \phi + \theta\eta$ in (7.1.17) to get

$$[\phi] - \frac{\sigma^+ + \sigma^-}{2}[\lambda] = -\theta[\eta] - \frac{1}{v}[q]$$

$$= \frac{\theta}{v}\left\{-v[\eta] - \left[\frac{q}{\theta}\right]\right\} \tag{7.1.19}$$

$$\leq 0,$$

where we have used (7.1.15) and $\theta = \theta^+ = \theta^-$. Then, (7.1.16) and (7.1.18) have nice interpretations in terms of the graph of $\sigma(\lambda, \theta)$, the isothermal stress–stretch curve at this temperature. It may look like Fig. 7.1.

From (7.1.16), the angle χ in the diagram satisfies

$$\tan \chi = \frac{[\sigma]}{[\lambda]} = \rho v^2, \tag{7.1.20}$$

---

[2]Our treatment is oversimplified. In dealing with moving phase boundaries, it is of some importance to account for surface, as well as bulk, energies associated with the interface between phases and workers consider possibilities other than (7.1.18). Three-dimensional theory of this kind is covered by Gurtin [35].

so we are considering χ to be a rather small angle. in (7.1.19), $[\phi]$ is the (signed) area under the stress–strain curve between λ^+ and λ^-, being negative in the illustration. Also, the (signed) area between λ^+ and λ^-, bounded above by the chord joining (σ^+, λ^+) to (σ^-, λ^-) is $(\sigma^+ + \sigma^-)[\lambda]/2$, which is also negative in the illustration. So, it is a matter of comparing these areas in deciding as to whether (7.1.19) is satisfied. In the situation pictured, it will be if λ^+ is close enough to the local maximum for the graph.

Impacts, etc., are likely to generate waves which travel much faster, being more like shock waves in air. Here, workers often assume that the effects of heat conduction can be neglected, using

$$q = 0 \qquad (7.1.21)$$

in place of (7.1.18). Then (7.1.17) reduces to what is often called the Rankine–Hugoniot equation

$$[\varepsilon] = \frac{\sigma^+ + \sigma^-}{2}[\lambda]. \qquad (7.1.22)$$

With an equation of the form $\varepsilon = \varepsilon(\lambda, \eta)$, this relates the two states $(\lambda^\pm, \eta^\pm)$ in a manner depending only on the material. Also, (7.1.15) reduces to

$$\eta^- \geq \eta^+. \qquad (7.1.23)$$

If this constitutive equation is known, one can select (λ^+, η^+) and, from (7.1.22) and (7.1.23), find all the thermodynamically admissible states (λ^-, η^-) which generally lie on a curve in the λ–η plane, the Rankine–Hugoniot curve. In a similar way, (7.1.16) generates a set of curves, one for each value of v, called Rayleigh curves. For a given value of v, the possible states behind correspond to points where this Rayleigh curve intersects the Rankine–Hugoniot curve when they satisfied (7.1.23). In this context, it is of interest to consider weak shocks, where $[\lambda]$, $[\eta]$, etc., are suitably small. Think of fixing (λ^+, η^+) and solving (7.1.22) for $[\eta]$ in terms of $[\lambda]$, assuming $[\eta]$ approaches zero as $[\lambda]$ does. A first approximation to v can be obtained from (7.1.16) by letting $[\lambda] \to 0$, $\eta^- \to \eta^+$ which gives

$$\rho v_0^2 = \frac{\partial \sigma}{\partial \lambda}(\lambda^+, \eta^+), \qquad (7.1.24)$$

v_0 being called the acoustic wave speed, measured by the slope of the isentropic stretch–strain curve at the values (λ^+, η^+) of interest. To estimate what (7.1.22) gives in this limit, use the linear approximation

$$\begin{aligned}
\varepsilon^- &= \varepsilon(\lambda^-, \eta^-) \\
&\cong \varepsilon(\lambda^+, \eta^+) + \frac{\partial \varepsilon^+}{\partial \lambda}(\lambda^- - \lambda^+) + \frac{\partial \varepsilon^+}{\partial \eta}(\eta^- - \eta^+) \\
&= \varepsilon^+ - \sigma^+[\lambda] - \theta^+[\eta],
\end{aligned}$$

so

$$\frac{[\varepsilon]}{[\lambda]} = \sigma^+ + \theta^+ \frac{[\eta]}{[\lambda]} = \frac{\sigma^+ + \sigma^-}{2}. \tag{7.1.25}$$

Letting $[\lambda] \to 0$ and $\sigma^- \to \sigma^+$, we get $[\eta]/[\lambda] \to 0$. So, an expansion of $[\eta]$ in powers of $[\lambda]$ gives zero for the linear term, suggesting it looks like

$$[\eta] = a[\lambda]^2 + b[\lambda]^3 + \dots , \tag{7.1.26}$$

Then, to second order in $[\lambda]$,

$$\varepsilon^- \cong \varepsilon^+ - \sigma^+[\lambda] - \theta^+ a[\lambda]^2 + \frac{1}{2}\left(\frac{\partial^2 \varepsilon}{\partial \lambda^2}\right)[\lambda]^2,$$

and, on the right side of (7.1.22), we can use a first-order approximation

$$\sigma^- = \frac{\partial \varepsilon(\eta^-, \lambda^-)}{\partial \lambda} \cong \sigma^+ + \left(\frac{\partial^2 \varepsilon}{\partial \lambda^2}\right)^+ (\lambda^- - \lambda^+).$$

Putting these approximations into (7.1.22), we get $a = 0$. Further calculation indicates that, in general, $b \neq 0$ so, in first approximation

$$[\eta] \propto [\lambda]^3, \tag{7.1.27}$$

for $[\lambda]$ small. It then follows that, to first order in $[\lambda]$,

$$[\sigma] = \rho v_0^2 [\lambda], \tag{7.1.28}$$

where v_0 is the acoustic speed, given by (7.1.24). One is then in the realm of linear theory and the pulses accurately described by this are often called acoustic waves or stress waves. To a good approximation equality then holds in (7.1.15). Accepting this, workers then do not worry about the implications of the possible inequality.

Bear in mind that here we only explored conditions relating to a discontinuity. Complete analysis of a problem also involves satisfying equations of the kind discussed in Chapter 2, where solutions are smooth. Also, we have emphasized how thermodynamic reasoning is used and other kinds of reasoning are employed. In practice, one encounters situations in which more than one solution satisfy all the conditions discussed. One of the kinds of reasoning used to try to decide which is best then involves considering the possibility that the shock can generate pulses of smaller amplitude. Consider the front side and suppose that these weaker waves have speeds faster than that of the shock. They will then tend to move out ahead of it. The notion is that this will tend to make the shock smooth out, almost instantaneously. So, the shock is more likely to persist if these weaker waves travel more slowly. Similarly, it is better if weaker waves on the back side travel more quickly than the shock. One can use (7.1.24) as a first estimate of the weak wave speed ahead, using the analogous estimate behind, to be compared with the shock speed, given by (7.1.16). Notice that this uses the slope of the isentropic stress–stretch curve. If one uses the slopes of the isothermal stress–stretch curve in Fig. 7.1 to estimate the weak wave speeds, the situation looks favourable at λ^-, unfavorable at λ^+. One could make it appear more favorable by increasing λ^+ sufficiently to make the shock chord tangent at λ^+. It is not difficult to calculate the difference between the isentropic and isothermal slopes at corresponding values of θ and η, so you can try this to see if you come to a different conclusion.

7.2   Breaking Bars

As was mentioned in the preceding section, it is sometimes relevant to consider surface energies associated with discontinuities. Let us consider a different possibility of this kind. We return to the equilibrium problems involving bars in hard loading devices, discussed in Section 3.2. Generally, this loading makes the relevant energy higher than it would be if the bar were unloaded. There is a way for this energy to be reduced. Let the bar break and this will certainly cause it to be unstressed, making $y(x)$ itself discontinuous. This is a possibility we have avoided considering, although we all know that it can happen. However, the reasoning seems to indicate that the slightest stress should suffice to break it, which does seem contrary to experience. Here, the reasoning commonly accepted is that the process of breaking creates new surfaces and with these is associated an energy proportional to their area. Usually, this is considered as the area occurring when the sample again becomes unstressed. So, if our bar breaks cleanly on a plane whose normal is parallel to the long direction, represented mathematically by our x-interval, this produces an energy

$$e = 2G\,A_R.$$

with A_R our reference area, G being the energy per unit reference area. The factor of 2 arises because the break obviously creates two surfaces. Solutions of this kind discussed in Section 3.2 then apply as long as F, given in (3.2.2), is smaller than e and, when $F > e$, the bar should break. There is an intermediate possibility that a crack will be produced, go part of the way through the specimen and stop. One then needs two- or three-dimensional theory to analyse the associated adjustment in the Helmholtz free energy. In some cases it diminishes enough to make it energetically disadvantageous for the crack surface area to increase. This provides some basis for understanding whether an expensive structure containing such flaws can be considered safe, among other things.

7.3   A Peeling Problem

Somewhat similar to the previously mentioned crack propagation problems are peeling problems. Suppose that a very thin and flexible adhesive tape has adhered to a flat surface of a stationary rigid body. We consider the possibility of peeling it off, by applying a tensile force as a dead load at one end. As usual, we assume that the ambient temperature is fixed. Think of our tape as a rectangular bar, with width a, thickness b, $b/a << 1$ and assume that it has been applied so that it is unstressed, although peeling it will change this. Then, for the bar, it is convenient to use the strain energy per unit volume $W(\lambda)$, at the prevailing temperature, giving

$$F_{\mathrm{B}} = ab \int_0^L W(\lambda)\,dx, \qquad W(1) = 0, \qquad (7.3.1)$$

as the relevant energetic contribution for it. If the force peels off part of the tape we will have a configuration like that pictured in Fig. 7.2.

In principle, there is some energy associated with bending the tape, which we neglect. Where the tape still adheres, we assume that the bar stays unstretched ($\lambda = 1$). In the remainder, there will be some other stretch λ, the distance d in the figure being given by

$$d = \int_{x_0}^L \lambda\,dx. \qquad (7.3.2)$$

There is the usual potential involved with the dead loading device. For it, we need the scalar product of the vector force with the vector displacement or, what is the same, f, multiplied by the component of displacement which is parallel to the force. Using simple geometry, this gives the potential

$$-f[d - (L - x_0)\cos\varphi]. \qquad (7.3.3)$$

Also, we introduce the idea that there is a surface energy to be added, proportional to the area occupied by the part of the tape which has come off, before it was removed. So, it will be of the form

$$e = G_a(L - x_0)a,$$

where G_a is a positive constant depending on the tape and the material from which the rigid body is made. It is a matter of experience that a given tape will adhere better to some materials than to others and it is here that we account for this. Adding the contributions we get the thermodynamic potential

$$F = ab \int_{x_0}^L W(\lambda)\,dx - f[d - (L - x_0)\cos\varphi] + G_a(L - x_0)a, \qquad (7.3.4)$$

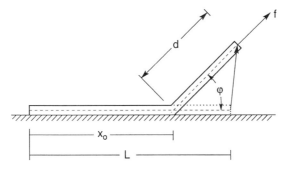

FIGURE 7.2. A fixed force of magnitude f, acting at the angle φ, peels part of the tape off the rigid body. The nearly vertical arrow indicates the displacement at the end.

with d given by (7.3.2). We will disregard the possibility that the tape breaks or that the rigid body gets damaged. We do need to bear in mind that the configuration pictured can hardly apply if the tape is all peeled off, or if none of it is. However, for the time being, we will not worry about this. For other values of x_0, we can vary x_0 and the bar deformation independently. Varying the latter gives, as an equilibrium equation, the force balance

$$abW'(\lambda) = ab\sigma = f. \qquad (7.3.5)$$

For simplicity, we assume that $W(\lambda)$ is convex, so this determines just one value of λ. Putting this back in (7.3.4) gives

$$F = (L - x_0)\{ab[W(\lambda) - (\lambda - \cos\varphi)\sigma] + aG_a\}. \qquad (7.3.6)$$

Clearly, this has a minimum at

$$x_0 = 0 \quad \text{when } ab[W(\lambda) - (\lambda - \cos\varphi)\sigma] + aG_a < 0 \qquad (7.3.7a)$$

$$x_0 = L \quad \text{when } ab[W(\lambda) - (\lambda - \cos\varphi)\sigma] + aG_a > 0 \qquad (7.3.7b)$$

and is independent of x_0 when

$$b[W(\lambda) - (\lambda - \cos\varphi)\sigma] + G_a = 0. \qquad (7.3.8)$$

Now, if no tape is peeled off, the obvious value of F is $F = 0$ and (7.3.6) reduces to this when $x_0 = L$. Then, (7.3.7b) gives a reasonable estimate of what is needed for such configurations to be stable. Similarly, (7.3.8) gives a critical condition for the tape to begin to peel off. Once the inequality in (7.3.7a) holds, it is better for x_0 to decrease, that is, more of the tape to peel off. Physically, (7.3.7a) cannot be trusted, implying that the tape can still support a stress when the area with which it contacts the rigid body reduces to zero. Instead, it will in fact lose its ability to be in equilibrium supporting the force applied at one end. To analyze what then happens, one would need to say more about the mechanism for applying the force and we will not pursue this. Otherwise, the model is reasonably consistent with experience. Let us try to obtain a better understanding of what it predicts. If, as was assumed, $W(\lambda)$ is a convex function of λ, $\sigma(\lambda)$ is a monotonically increasing function of λ, with $\sigma(1) = 0$. Thus the values of λ satisfying (7.3.5) will, for positive f, increase with f and be greater than one. Consider, for any fixed value of φ,

$$g(\lambda) = ab[W(\lambda) - (\lambda - \cos\varphi)\sigma] + aG_a, \qquad (7.3.9)$$

we have

$$g(1) = aG_a > 0, \qquad (7.3.10)$$

and

$$\begin{aligned} g'(\lambda) &= ab[W'(\lambda) - \sigma - (\lambda - \cos\varphi)\sigma'] \\ &= -ab(\lambda - \cos\varphi)\sigma' < 0. \end{aligned} \qquad (7.3.11)$$

For $\lambda > 1$ and $(\lambda - 1)$ sufficiently small, $g(\lambda)$ will be positive by continuity and this will correspond to having f small enough. Then, (7.3.7b) implies that the whole tape should remain in contact with the other body. As f increases, (7.3.5) implies that $g(\lambda)$ decreases. Mathematically, it could approach a positive constant as λ becomes large. Physically, something must happen if the force gets sufficiently large but there is some possibility that this may occur, by breaking out chunks of the other body, for example. Assuming g does continue to decrease, it will vanish at some critical value f_c of the force. Here, in principle, any part of the tape could peel off with the rest staying in contact. For $f > f_c$ we will have $g < 0$, the somewhat shaky (7.3.7a) seeming to imply that the whole tape should come loose. With the inevitable experimental errors, it is unlikely that we will attain the value of f_c exactly. Granted this, one should see the tape either staying in complete contact, or coming completely free, depending on the size of the force, in this kind of experiment. This is consistent with experience on relatively long tapes, for which L/a is relatively large. Generally, applying a force in the manner indicated either leaves the whole tape in contact with the other material, or the whole tape comes off. If a force induces it to come off, a larger force simply makes it peel off more quickly, assuming that there are no other complications such as having the tape break. Also, from our considerations, the critical force f_c occurs when (7.3.8) holds and this clearly depends on the angle φ, which should be in the range $0 \leq \varphi \leq \pi$, physically. In this range and with $\sigma > 0$, as it must be, it is easy to see that for any fixed value of $\lambda > 1$, g increases with φ. Combine this with (7.3.11) and consider the critical values of λ, say $\lambda_c(\varphi)$, obtained by solving

$$g(\lambda_c, \varphi) = 0, \qquad (7.3.12)$$

and you can conclude that λ_c must decrease as φ increases. From this, it follows that f_c is a monotonically decreasing function of φ. At least qualitatively, this is also in accord with experience.

Another picture can be useful. Rewrite (7.3.8) as

$$(\lambda_c - \cos\varphi)\sigma(\lambda_c) - W(\lambda_c) = G_a/b. \tag{7.3.13}$$

Then, consider the graph of $\sigma(\lambda)$ in the λ–σ plane. The first term can be interpreted as the area of a rectangle with height $\sigma(\lambda_c)$, base $(\lambda_c - \cos\varphi)$, the length of a line segment on the λ axis, running from $\lambda = \cos\varphi < 1$ to $\lambda = \lambda_c > 1$. From this, we subtract $W(\lambda_c)$ which we know is the area under the stress–stretch curve between $\lambda = 1$ and $\lambda = \lambda_c$. Thus, the left side of (7.3.13) represents the area of the rectangle which remains after removing the latter part. When this area matches the number on the right, the equation is satisfied. From our previous considerations, there cannot be two values of λ_c satisfying it if $W(\lambda)$ is convex. If it is not, then one can get into complications such as are discussed in Chapter 3.

7.4   Another Peeling Problem

Here, we consider a procedure permitting us to peel off only part of the tape, also indicating how we might apply the tape so that, after it is placed, it will be unstressed. Again, we gloss over some physical difficulties occurring when only a small part near one end is in contact.

Here, the end $x = L$ is to be moved perpendicular to the rigid body to a height h, then held at this fixed position. Otherwise, the assumptions are similar to those made before. Figure 7.3 indicates the configuration to be considered.

Here, the angle φ is not given. As before, $W(\lambda)$ is assumed to be a convex function for simplicity and we again ignore complications associated with damage to the materials. It is then rather clear from our past experience that, in the part pulled off, the stretch λ will be constant, so we will assume this although it is something that could be proven. Here, with the end held fixed, no work will be done on it so, in place of (7.3.4), we now have

$$F = ab \int_{x_0}^{L} W(\lambda)\,dx + G_a(L - x_0)a \tag{7.4.1}$$

$$= a[bW(\lambda) + G_a]$$

where

$$l = L - x_0, \tag{7.4.2}$$

and note, from Fig. 7.3, that

$$d^2 = l^2 + h^2 = \lambda^2 l^2, \tag{7.4.3}$$

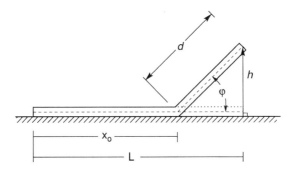

FIGURE 7.3. Raising the end of the tape vertically, to a height h, partially peels off the tape.

and

$$\tan \varphi = h/l. \tag{7.4.4}$$

Thus, two of the three variables (λ, l, φ) can be expressed in terms of the third. We could take any one as the basic independent variable, it being a matter of judgment which is best. First, let us try using

$$\mu = l/h \Rightarrow 0 \leq \mu \leq L/h. \tag{7.4.5}$$

Then, (7.4.3) gives

$$\lambda = \sqrt{1 + \mu^{-2}} > 1, \tag{7.4.6}$$

and (7.4.1) reduces to the form

$$F/ah = U(\mu) + \mu G_a, \tag{7.4.7}$$

where

$$U(\mu) = b\mu W\left(\sqrt{1 + \mu^{-2}}\right). \tag{7.4.8}$$

If μ is very small, λ is very large, approaching infinity as $\mu \to 0$. Physically, it is then unlikely that we will have an end point minimum at $\mu = 0$. Also, having μ at the other limit really means that all the tape has peeled off. If we continue to hold one end fixed the remainder will be subject to no force, so no work will be done on the tape. It should then come to equilibrium, unstressed, giving

$$F = aLG_a \tag{7.4.9}$$

as a reasonable estimate of the energy it will then have. The remaining possibility is to have equilibrium with μ not at either limit. Then, mathematically, analysis of (7.4.7) is essentially the same as that of analyzing bars under dead loads discussed in Section 3.1, or of the balloons under fixed pressure discussed in Section 5.1. The equilibrium equation is

$$U'(\mu) = -G_a, \tag{7.4.10}$$

the second derivative test for stability giving

$$U''(\mu) \geq 0. \tag{7.4.11}$$

Conditions for stable equilibria can be pictured, as before, in terms of the graph of $U(\mu)$ or of $U'(\mu)$. Some points are worth noting. Suppose (7.4.10) is satisfied by some value of $\mu = \bar{\mu}$. This is acceptable only if it satisfies $\bar{\mu} < L/h$, which will be true if we make h small enough, but not if h is too large. For the range of h for which it is acceptable, (7.4.4) will give a value of $\varphi = \bar{\varphi}$ which is independent of h. Ignoring the question of stability, the implication is that if we increase h a little, more of the tape will come loose, enough to get $\bar{\varphi}$ back to its original value with no change in the value of λ. Clearly, it can so adjust only as long as there remains enough tape to be peeled off, that is for

$$h \leq L \tan \bar{\varphi}. \tag{7.4.12}$$

For such configurations to be more stable than they would be if the tape pulled off completely, using (7.4.7) and (7.4.9) gives the condition

$$h[U(\bar{\mu}) + \bar{\mu} G_a] < L G_a. \tag{7.4.13}$$

Now consider (7.4.8). With (7.4.6), straightforward calculations give

$$U'(\mu) = b\left[W(\lambda) + \mu \frac{d\lambda}{d\mu} \sigma(\lambda)\right]$$

$$= b\left[W(\lambda) - \left(\mu^{-2}\sqrt{1 + \mu^{-2}}\right)^{-1} \sigma(\lambda)\right] \tag{7.4.14}$$

$$= b\left[W(\lambda) - \left(\lambda - \frac{1}{\lambda}\right)\sigma(\lambda)\right] = -G_a.$$

Note first that, from Fig. 7.3,

$$\cos\varphi = \frac{L - x_0}{d} = \frac{l}{d} = \frac{1}{\lambda}. \tag{7.4.15}$$

With this, we see that (7.4.14) agrees with (7.3.8), which is not surprising. Also

$$U'' = \frac{d\lambda}{d\mu}\frac{dU'}{d\lambda} = \frac{d\lambda}{d\mu} b\left[\sigma(\lambda)\left(1 + \frac{1}{\lambda^2}\right) - \left(\lambda - \frac{1}{\lambda}\right)\sigma'(\lambda)\right]$$

$$= -\frac{d\lambda}{d\mu} b\left[\frac{1}{\lambda^2}\sigma(\lambda) + \left(\lambda - \frac{1}{\lambda}\right)\sigma'(\lambda)\right]. \tag{7.4.16}$$

Now, from (7.4.6), $d\lambda/d\mu < 0$ and $\lambda > 1$. From our assumption that $W(\lambda)$ is convex with $\sigma(1) = 0$, it follows that $\sigma(\lambda) > 0$ for $\lambda > 1$ and $\sigma(\lambda) \geq 0$, from which

$$U''(\mu) > 0. \tag{7.4.17}$$

There then can be no more than one equilibrium value $\bar{\mu}$ of the kind discussed above. Also there can be at most one value of the angle $\bar{\varphi}$, in the range $0 \le \bar{\varphi} \le \pi$. Assume there is one. Then, for h satisfying (7.4.11), we have equilibrium configurations satisfying all the requirements for stability except perhaps (7.4.12). With $\bar{\mu}$ a fixed number, this will also hold if h is sufficiently small. For the largest possible value of h, given by the equality in (7.4.11), it is easy to see that (7.4.12) fails to hold. Thus the two energies become equal at some value h_c, given by

$$ h_c = L \left[\frac{G_a}{U(\bar{\mu}) + \bar{\mu}G_a} \right]. \tag{7.4.18} $$

So for $h < h_c$, the more stable configuration has the tape partially adhering, enough to conform to the angle $\bar{\varphi}$. For $h > h_c$, the more stable configuration changes, to have the tape completely removed. From the analysis, the possibility of the tape adhering when $h > h_c$ giving a metastable configuration does not seem unreasonable in cases where we are increasing h, to peel off more of the tape. In laying a tape down, we commonly hold the part near an end in place to get the process started. Clearly, our analysis does not cover this. Essentially, the theory is designed to apply to situations such that one can lay down and peel off the tape without damaging the tape or the other material.

To determine $W(\lambda)$ for $\lambda > 1$, most would try a simple tension experiment. Either of the two peeling experiments could be used to try to estimate G_a. Here, one could measure the stretch in the stretched part. Similarly, h_c might be measured. Think a little more about the measurables in the two experiments and you have a design for an experimental programme to provide a test of the theory in order to see if its predictions agree with the experimental findings.

This is a sample of ideas used in tackling problems relating to the adhesion of one material to another. Works dealing with this general topic include the books by Cheng [36] and Wu [37].

7.5   Exercises

For Exercises 7.1–7.3, adapt the theory of fast shock waves in bars to the theory of shearing of plates, using the constitutive equation

$$ \phi = \frac{a(1 - b\theta)\gamma^2}{2} - c \left[\theta \ln \left(\frac{\theta}{d} \right) - \theta + d \right], $$

where a, b, c, and d are positive constants. For various metals, empirical estimates of b give it the value $1/(2\theta_M)$, where θ_M is the melting temperature, so the theory does not apply if $2b\theta \ge 1$. If some prediction violates this, note it. For the exercises indicated, assume that, on one side of a shock, $\gamma = 0$, $\theta = d$, and on the other, $\gamma = \bar{\gamma} > 0$.

7.1. Derive the Rayleigh equation.

7.2. Derive the Rankine–Hugoniot equation.

7.3. What can you say about the admissibility of such a wave?

7.4. Consider the kind of peeling problem discussed in Section 7.2 for a linear elastic tape with the strain energy function $W = E(\lambda - 1)^2/2$, E being Young's modulus, a positive constant. Let $f(\psi)$ denote the lower bound of the force required to peel off the tape, when it is applied at the angle ψ. Derive formulae for the ratios

$$f\left(\frac{\pi}{2}\right)\bigg/f\left(\frac{\pi}{4}\right) \quad \text{and} \quad f\left(\frac{3\pi}{4}\right)\bigg/f\left(\frac{\pi}{4}\right).$$

7.5. Derive an equation relating isothermal and isentropic acoustic wave speeds for bars, in terms of things either measured or easily calculated from things measured in the common thermodynamic experiments.

8
Mixture Theory

8.1 General Remarks

From a macroscopic point of view, the composition of matter can change in various ways. It is a familiar fact that changes in the humidity occur in the air we breathe, causing wood to shrink or swell. This makes it reasonable to think that the wood absorbs different amounts of the vapour depending on its environment, producing an effect somewhat like thermal expansion. Materials like glass are clearly more reluctant to absorb the water, making it clear that the water vapour is not uniformly distributed in a room containing various kinds of materials, although it may be uniformly distributed within a particular substance.

Again, it is a matter of experience that different amounts of a solid can be dissolved in a given amount of a suitable fluid and that, if we change the temperature, say, some of the dissolved solid can come out of solution. Sometimes we use this to grow larger crystals from small ones. Not so different are the procedures used by a metallurgist to make alloys; that is, melt and mix together different kinds of solids, then cool the mix to solidify it. With the same ingredients mixed in different proportions he can obtain different alloys. For example, he might produce α or β brass by using different proportions of copper and zinc.

In many situations like this, it is reasonable to think that the ingredients retain their identity so the total mass of one remains fixed. We are then concerned with mixtures which are, in a sense, nonreacting. Chemical reactions can change the mass of particular ingredients with, for example, hydrogen and oxygen gases reacting to produce some water. Stoichiometry provides rules, restricting how the different masses can change in such reactions.

Gibbs' general ideas of thermodynamic equilibrium and stability can be applied to such phenomena. Indeed, Gibbs was a pioneer in developing such theory. Here, we will discuss the most elementary format for nonreacting mixtures, in situations where effects of shear stresses in the solids can be ignored.

To deal with such questions one needs to think of equations describing parts of a system and this involves some judgment. In the first example mentioned, most would think of the air and water vapour as one material, described by one constitutive equation. A different equation would be used for the wood and its water vapour. Generally, one idea is to try to arrange that each constitutive equation can reasonably be considered smooth enough to let us use common tools of analysis in each part. As we have seen before, smooth constitutive equations can produce equilibria with discontinuities and we tend to judge things on the basis of observations of equilibria. Given this, it is inevitable that we exercise some judgment. In the simplest situations, one can reasonably decide how to further subdivide so equilibria are also smooth in each part.

8.2   Elementary Theory

We consider a hypothetical example to introduce some of the ideas. Picture a mixture filling some volume V. Consider it to be composed of homogeneous parts. That is, within a part, relevant state variables, such as temperature, etc., can be considered to be independent of position. So the presumption is that we know how to subdivide so that the equilibria are smooth in each part. The parts will occupy volumes

$$V_i, \qquad i = 1, \ldots, m$$

with

$$\sum_{i=1}^{m} V_i = V. \tag{8.2.1}$$

An identifiable ingredient, such as our water vapour, may be present in all of these parts and we allow that this can happen, initially, and that the amount in one part can be considered to change. Suppose that we have n ingredients, with masses M_α, $\alpha = 1, \ldots, n$. Then, if $M_\alpha^{(i)}$ is the mass of the αth ingredient in the volume V_i, we must have

$$\sum_{i=1}^{m} M_\alpha^{(i)} = M_\alpha, \qquad \alpha = 1, \ldots, n, \tag{8.2.2}$$

it being possible that some of the $M_\alpha^{(i)}$ vanish. This is a static theory, so we are not concerned with motion. With the ith part we associate a total energy E_i and entropy S_i and a constitutive equation. That for $i = 1$ is assumed to be a smooth function of the form

$$E_1 = E_1(V_1, S_1, M_1^{(1)}, \ldots, M_n^{(1)}), \tag{8.2.3}$$

with similar assumptions for E_2, etc. It is possible that one constitutive equation applies to two more parts. The assumptions are fairly reasonable if we have solid parts separated by fluid parts, so the solids can adjust their shape fairly freely to avoid shear stresses. The idea is that the partial volumes and masses can vary, subject to the requirement that (8.2.1) and (8.2.2) be satisfied; we do not want overlapping volumes, for example. We set

$$p_i = -\frac{\partial E_i}{\partial V_i}, \tag{8.2.4}$$

interpreted as the pressure in the ith part,

$$\theta_i = \frac{\partial E_i}{\partial S_i}, \tag{8.2.5}$$

the temperature in this part and

$$\mu_{i\alpha} = \frac{\partial E_i}{\partial M_\alpha^{(i)}}, \tag{8.2.6}$$

these being called *chemical potentials*. Now look at the differential of E_1. One term, $-p_1\,dV_1$ can reasonably be interpreted as the work done by the pressure as the volume V_1 undergoes an incremental change. Another, $\theta_1\,dS_1$, similarly can be identified with heat supplied reversibly. If we stopped here, things would make reasonable sense in terms of the first law. However, there is a problem in dealing with the changes in E_1 produced by changing the masses in V_1 which does not really fit the first law. One could try to generalize the first law to cover this. What is more often done, sometimes, tacitly, is to consider that any thermodynamic system deals with a fixed set of matter. From this viewpoint the energy E_1 is then an energy associated with part of a thermodynamic system, which is not itself a thermodynamic system. The usual idea that energies and entropies are additive then gives for the energy E and entropy S of the system,

$$E = \sum_{i=1}^{m} E_i,$$

$$S = \sum_{i=1}^{m} S_i. \tag{8.2.7}$$

Physically, we can reasonably think of this as a mechanically isolated system if we think of V as the volume of a fixed region so that forces acting on its boundary do no work. Further, it should prevent any matter from leaving or entering, so that M_α become fixed constants. Thermally, a likely assumption is that the system is in contact with a heat bath at the constant temperature θ_B. Then, the appropriate thermodynamic potential is the ballistic free energy,

$$E_B = E - \theta_B S. \qquad (8.2.8)$$

In considering possible variations, one needs to consider the possibility that some of the masses $M_\alpha^{(i)}$ or some of the volumes may vanish.[1] This reflects the experience that, in the first case, some materials are very reluctant to absorb others. In the second case, we know that a solid lump is sometimes completely dissolved in a fluid. The fluid might well be capable of dissolving more if it were supplied. Clearly, we cannot allow variations making masses or volumes negative. One can modify the scheme a little. For example, if one is convinced that it is quite impossible for one substance to penetrate another, one can impose the constraint that one of the $M_\alpha^{(i)}$ always vanishes. Materials like this are certainly of some importance, as containers for other materials, the thin films which serve as barriers to water but not air, etc. However, we will not consider this kind of possibility. It is assumed that

$$V_i = 0 \Rightarrow E_i = 0.$$

As usual, we proceed to use the first derivative test, to obtain equations of equilibrium. That is, we want the differential condition

$$dE_B = dE - \theta_B \, dS = \sum_{i=1}^{m} (dE_i - \theta_B \, dS_i) \geq 0, \qquad (8.2.9)$$

for all possible variations in the independent variables involved. There are the constraints,

$$dM_\alpha = \sum_{i=1}^{m} dM_\alpha^{(i)} = 0$$

$$\qquad\qquad\qquad\qquad\qquad (8.2.10)$$

$$dV = \sum_{i=1}^{m} dV_i = 0,$$

---

[1] In the case where the volume occupied by an ingredient approaches zero, it can be tricky to try to take limits of this kind. Usually, one can explore equilibria with or without complete dissolving, then determine which is most stable by an energy comparison. Think back to what we did for the balloon with pressure controlled.

at least. First consider varying only the entropies S_i. With (8.2.5), this gives, assuming no $V_i = 0$,

$$\sum_{i=1}^{m} (\theta_i - \theta_B)\, dS_i = 0.$$

With the usual assumption that entropies can be varied arbitrarily, this gives

$$\theta_i = \frac{\partial E_i}{\partial S_i} = \theta_B. \tag{8.2.11}$$

If some $V_i = 0$, one simply does not get the corresponding equation. Physically, in equilibrium the temperature throughout the system must reduce to that of the heat bath. As in other cases discussed before, most workers are willing to assume this and that (8.2.5) can be solved for S_i as a function of θ_i and the other variables. Then, one can replace E_i by the Helmholtz energy

$$F_i = F_i(V_i, \theta_B, M_\alpha^{(i)}) = E_i - \theta_B S_i. \tag{8.2.12}$$

By an exercise in calculus, one can then show that

$$p_i = -\frac{\partial F_i}{\partial V_i}, \qquad S_i = -\frac{\partial F_i}{\partial \theta_B}, \qquad \mu_{i\alpha} = \frac{\partial F_i}{\partial M_\alpha^{(i)}}. \tag{8.2.13}$$

Similarly, (8.2.9) can be replaced by

$$dF = \sum_{i=1}^{m} dF_i \geq 0. \tag{8.2.14}$$

So, let us use this formulation. First, consider possible equilibria for which none of the masses or volumes vanish, assuming that all the differentials satisfying (8.2.10) are permissible. With our assumption it is easy to see that, in (8.2.14), it is not possible for the inequality to hold. In part, we then get the conditions

$$\sum_{i=1}^{m} \frac{\partial F_i}{\partial V_i}\, dV_i = 0$$

or, with (8.2.13)

$$\sum_{i=1}^{m} p_i\, dV_i = 0, \tag{8.2.15}$$

for all dV_i satisfying (8.2.10). Clearly, we can assign arbitrary values to $dV_2, \ldots, dV_m$ and solve (8.2.10) for dV_1. If we multiply the sum in (8.2.10) by p_1 and subtract this from (8.2.15), we get

$$\sum_{i=2}^{m} (p_i - p_1)\, dV_i = 0,$$

where the dV_i occurring here are arbitrary, so we must have

$$p_1 = p_2 = \cdots = p_m, \qquad (8.2.16)$$

the equality of pressures which we might also infer, mechanically, as a balance of forces. Similar consideration of variations of masses gives, with (8.2.13), the equality of chemical potentials given by

$$\mu_{1\alpha} = \mu_{2\alpha} = \cdots = \mu_{m\alpha}, \qquad \alpha = 1, \ldots, n. \qquad (8.2.17)$$

So, (8.2.16) and (8.2.17) give the equations of equilibrium. Of course, we also must satisfy (8.2.1) and (8.2.2). With M_α and V here considered as given, the number of equations is then the same as the number of unknowns (V_i and $M_\alpha^{(i)}$).

While we shall not discuss in detail the equilibrium conditions which apply when some of the volumes and/or masses vanish, there is another assumption concerning constitutive equations which deserves to be mentioned. Briefly, it is that, if we multiply all masses and volumes by a common factor, we can get the new energy by multiplying the old value by the same factor. Roughly, it is the proportions which matter, not the absolute amounts. For example, for the function F_1, we assume that for any positive number k, and any possible values of the arguments, we have

$$F_1(kV_1, \theta, kM_\alpha^{(1)}) = kF_1(V_1, \theta, M_\alpha^{(1)}). \qquad (8.2.18)$$

as a restriction on the form of this function. If we differentiate this relation with respect to k, then set $k = 1$, we obtain the identity

$$-p_1 V_1 + \sum_{\alpha=1}^{n} \mu_{1\alpha} M_\alpha^{(1)} = F_1. \qquad (8.2.19)$$

There are various ways of taking care of this restriction. For example, taking $k = 1/V_1$ in (8.2.18) and setting

$$F_1(1, \theta, \rho_\alpha^{(1)}) = \phi_1(\theta, \rho_\alpha^{(1)}), \qquad \rho_\alpha^{(1)} = M_\alpha^{(1)}/V_1,$$

we have

$$F_1 = V_1 \phi_1(\theta, \rho_\alpha^{(1)}), \qquad (8.2.20)$$

where $\rho_\alpha^{(1)}$ are interpretable as mass densities, ϕ as the Helmholtz free energy per unit volume. As is easy to check, if we take any function ϕ_1 of the indicated arguments and use (8.2.20) to define F_1, it will satisfy (8.2.18) and (8.2.19). This opens the door for dealing with problems in which the mass densities $\rho_\alpha^{(1)}$ might vary with position, as they will in a gravitational field. This can be important for estimating water content in soil near a deep lake, for example. Then, the total Helmholtz free energy F_1 in a region occupied by this component could be taken as

$$F_1 = \int \phi_1(\theta, \rho_\alpha^{(1)}) \, dV, \qquad (8.2.21)$$

giving (8.2.20) when the integrand is constant and the volume of the region is V_1. Clearly, much the same considerations apply to the other F_i and to the E_i, with the understanding that entropies are to be replaced by entropies per unit volume. With (8.2.20), one can verify that (8.2.19) reduces to

$$-p + \sum_{\alpha=1}^{n} \mu_{1\alpha} \rho_{\alpha}^{(1)} = \phi_1. \tag{8.2.22}$$

A slightly different formulation can be obtained by introducing

$$m^{(1)} = \sum_{\alpha=1}^{n} M_{\alpha}^{(1)}, \tag{8.2.23}$$

the total mass in V_1 and the concentrations (mass fractions)

$$c_{\alpha}^{(1)} = M_{\alpha}^{(1)}/m^{(1)}, \tag{8.2.24}$$

numbers lying between zero and one, satisfying

$$\sum_{\alpha=1}^{n} c_{\alpha}^{(1)} = 1. \tag{8.2.25}$$

Then, one can introduce the specific volume

$$v_1 = V_1/m^{(1)} = \frac{1}{\rho_1}, \tag{8.2.26}$$

where ρ_1 is the total mass density in V_1. Then, taking $k = 1/m^{(1)}$, we see that we can also write F_1 in the form

$$F_1 = m^{(1)}\psi(v_1, \theta, c_{\alpha}^{(1)}) = V_1 \rho_1 \psi(v_1, \theta, c_{\alpha}^{(1)}), \tag{8.2.27}$$

and we can, if we like, use (8.2.25) to eliminate one of the $c_{\alpha}^{(1)}$.

Originally, Gibbs introduced this kind of theory as a theory of fluid mixtures but later workers, like those interested in alloys, have applied it to cases where some of the parts are solid. Confining solids to a fixed region can easily induce shear stresses, so it is a little more natural to think of V as the volume of a region partly occupied by solids which are surrounded by a fluid at a fixed pressure p, where the shape and volume V of the region can change, to help avoid such complications. Then, the surrounding fluid serves as a loading device which can do work on the solids. This changes the relevant thermodynamic potential from F to

$$G = F + pV, \tag{8.2.28}$$

sometimes called the Gibbs' function. Involved, tacitly, is the assumption that the solids will stay in V, and not diffuse into the surrounding fluid, an assumption which can be quite good, or very bad, depending on the materials involved. It is a poor assumption for dry ice in air at room temperature, for example. Here, the subvolumes are still considered to add up to V, as indicated by (8.2.1), but V can vary. Also, (8.2.2) should hold with M_α fixed. Again assuming non-zero volumes and masses, we get, as equilibrium equations,

$$p_1 = \cdots = p_m = p, \tag{8.2.29}$$

as a replacement for (8.2.16) and, as before, (8.2.17), the equality of chemical potentials. Here, V is not known but p is regarded as a given constant.

Often, workers will go a step further. Suppose p_1 is a monotonically decreasing function of V_1 when the other variables are held fixed ($\partial p_1 / \partial V_1 = -\partial^2 F_1 / \partial V_1^2 < 0$). Then, we can solve $p_1 = p$ for $V_1 = f(p, \theta, M_\alpha^{(1)})$, and express

$$G_1 = F_1 + pV_1 \tag{8.2.30}$$

as a function of these variables. Then, an exercise in calculus gives

$$\frac{\partial G_1}{\partial \theta} = -S_1, \qquad \frac{\partial G_1}{\partial p} = V_1, \qquad \frac{\partial G_1}{\partial M_\alpha^{(1)}} = \mu_{1\alpha}. \tag{8.2.31}$$

With similar assumptions, we can do the same for the other ingredients and get

$$G = \sum_{i=1}^{m} G_i = \sum_{i=1}^{m} (F_i + pV_i) = F + pV,$$

$$= G(\theta, p, M_\alpha^{(i)}). \tag{8.2.32}$$

Breakdowns in the assumed invertibility associated with phase transitions sometimes occur. Attempts to perform the inversion can then lead to Gibbs' functions which are multivalued and/or exhibit singularities when the corresponding Helmholtz free energy function is well-behaved. This is not to say that the latter cannot be ill-behaved, but the situation described is rather common. Commonly, phase diagrams are drawn to indicate under what conditions phase transformations take place. Typical diagrams for alloys are discussed by Ricci [38].

After taking care of satisfying the equilibrium equations one would like to test for stability and a common procedure is to try to use the second derivative test. If, say, one is using (8.2.9) and is not concerned with the possibility of zero masses or volumes, one would calculate the second differential of E_B, a quadratic in the differentials of $M_\alpha^{(i)}$, V_i and S_i. Some of the differentials can be eliminated, expressed as linear combinations of others, to get a quadratic form involving differentials which can be varied arbitrarily. For metastability one would like this to be positive or maybe zero, for all choices of the differentials. Schematically, one has a quadratic condition like

$$Q = \sum_{(i,j=1)}^{n} a_{ij}x_ix_j \geq 0 \tag{8.2.33}$$

for all choices of x_i, and we can assume that

$$a_{ij} = a_{ji}.$$

If this is not true, we can replace the coefficients a_{ij} by $(a_{ij} + a_{ji})/2$, which does not affect Q. One way of proceeding is to determine all of the eigenvalues of the matrix $\| a_{ij} \|$; they must be nonnegative. In most cases it is easier to use another test. Suppose first that the inequality is to be satisfied in the strict sense, $Q = 0$ only when all $x_i = 0$. Then, in the matrix $\| a_{ij} \|$, go down the main diagonal, blocking out 1×1, $2 \times 2, \ldots$, matrices as indicated by the diagram

$$\left\| \begin{array}{c|c|c|c} a_{11} & a_{12} & a_{12} & a_{1n} \\ a_{12} & a_{22} & a_{32} & a_{2n} \\ a_{13} & a_{32} & a_{33} & a_{3n} \\ & & & | \\ | & & & | \\ a_{1n} & & & a_{nn} \end{array} \right\|.$$

Then, we will have $Q > 0$ if, and only if, the determinants of all these matrices are positive.[2] This gives a string of inequalities such as

$$a_{11} > 0, \qquad a_{11}a_{22} - a_{12}^2 > 0, \ldots, \tag{8.2.34}$$

not so difficult to check.

If (8.2.33) does not hold in the strict sense, there will be some nonzero x_i for which $Q = 0$ and, with $Q \geq 0$, these will satisfy the first derivative test for a minimum,

$$\sum_{j=1}^{n} a_{ij}x_j = 0, \tag{8.2.35}$$

linear equations which are fairly easy to solve. One can then make a change of variables to get a quadratic form in fewer variables, which is strictly positive, then use the above criterion for it.

This covers some of the basic ideas used in formulating elementary problems involving mixtures, enough to indicate that they are similar to ideas we have used before. As might be expected from our discussion, analysis of physical problems of this kind tends to be rather complicated but involves ideas much like those used before.

---

[2] A proof is given in Frazer et al. [39].

In part, we introduced this to cover stability problems which can involve a large number of variables. One can encounter these in various other situations. Even with our simple balloon problems, one encounters the need to take several variables into account if one considers many balloons, connected by pipes so air can move from one to the others. In part, we have also introduced this to indicate that, while deformations and temperatures generally have some relevance to the description of the states of solids, other quite different variables may also be needed. Amongst other things, this indicates that the "typical" thermodynamic experiments discussed in Section 2.5 need to be revised to provide relevant information concerning the other variables. Finally, mixture theory is being used, increasingly, to help us understand a variety of phenomena, so it seems worthwhile to discuss it a little.

8.3 A Solid in an Ideal Gas

Here, we will consider one of the simplest situations employing two volumes, involving a solid surrounded by an ideal gas. The assumptions are as before, except that we assume that the solid remains intact, no part of it entering the surrounding gas. However, gas can move into the solid. The problem is to design the analogue of the thermodynamic experiments for bars discussed in Section 2.5, to determine F_1. If the index 1 refers to the solid, so it occupies the volume V_1 etc., our assumption is that

$$M_1^{(2)} = 0 \Rightarrow M_1^{(1)} = M_1, \tag{8.3.1}$$

a fixed constant for a given sample of the solid. For the solid, the Helmholtz free energy function is an unknown function of the form

$$F_1(V_1, M_1, M_2^{(1)}, \theta), \tag{8.3.2}$$

and we are concerned with this at a fixed values of θ, in considering isothermal mechanical experiments. This leaves V_1 and $M_2^{(1)}$ as two variables which we want to vary independently. For the surrounding gas, (8.3.1) permits us to continue to model this as an ideal gas, so we will have

$$p_2 = \frac{R\theta M_2^{(2)}}{V_2}. \tag{8.3.3}$$

Also, the usual assumption is that C_v, the specific heat at constant volume, is constant. With this, the corresponding Helmholtz free energy function is of the form

$$F_2 = M_2^{(2)} \left[-R\theta \ln \frac{V_2}{M_2^{(2)}} - C_v \theta \ln \theta + a\theta + b \right], \tag{8.3.4}$$

where a and b are constants. This fits the format indicated by (5.1.7) in our discussion of balloons. In such cases, where the mass of gas is held fixed, the values we assign to a and b are unimportant. Here, their values will affect values of chemical potentials, so it seems safer to avoid assigning values to them to see what happens. To simplify notation, we write

$$M_2 = M, \qquad M_2^{(1)} = m \Rightarrow M_2^{(2)} = M - m. \qquad (8.3.5)$$

It is a reasonable guess that, to vary V_1 and m independently, we will need to use two control parameters. For one of these, let us try using the total volume V, with

$$V = V_1 + V_2, \qquad (8.3.6)$$

Physically, this could be done using a sturdy pressure chamber involving a movable piston as a wall. As the other, we consider using the amount of gas put into it, as measured by M. Physically, one wants to keep the walls from enough contact with the solid to induce shear stress, etc., so cases where $V_2 = 0$ are not of all interest. Said differently, we can vary V_2 sufficiently to get equality of pressures

$$p_2 = p_1 = -\frac{\partial F_2}{\partial V_1}. \qquad (8.3.7)$$

Otherwise, equilibria of the endpoint type could occur, with $m = M$ or $m = 0$ but, for the moment, we will exclude these cases. Then, we should have the equality of chemical potentials indicated by (8.2.13) and (8.2.17). With (8.3.4) and (8.3.5), we thus get

$$-\frac{\partial F_1}{\partial m} = R\theta \left[1 - \ln \left(\frac{V - V_1}{M - m} \right) \right] - C_v\theta \ln \theta + a\theta + b. \qquad (8.3.8)$$

Similarly, (8.3.3) and (8.3.5) give

$$\frac{\partial F_1}{\partial V_1} = -R\theta \left(\frac{M - m}{V - V_1} \right). \qquad (8.3.9)$$

So, for various values of θ, we should vary M and V as much as is feasible, performing the measurements required to obtain corresponding values of m and V_1. If all goes well, we then get empirical relations of the form

$$m = m(M, V, \theta), \qquad V_1 = V_1(M, V, \theta),$$

invertible to give relations of the form

$$M = M(m, V_1, \theta), \qquad V = V(m, V_1, \theta). \qquad (8.3.10)$$

Realistically, one might encounter complications more or less like those we have encountered in other situations, but we will ignore this.

Now consider what happens if we let M approach zero, which will also force m to approach zero. Since we are dealing with a solid, V_1 is not likely to approach V if we make V large enough, so the pressure should approach zero. Under these conditions, if we vary θ, the pressure will remain zero and the volume V_1 can be expected to change. This is the analogue of the thermal expansion discussed in Section 2.5. This gives a function which can be measured

$$V_1 = v(\theta), \quad \text{when } M = m = 0, \tag{8.3.11}$$

such that the derivative in (8.3.8) vanishes. With this, and (8.2.13), we then have

$$dF_1 = -\overline{S}_1\, d\theta, \tag{8.3.12}$$

where

$$\overline{S}_1 = S_1 \bigg|_{\substack{M=m=0 \\ V_1=v(\theta)}} = \overline{S}_1(\theta) \tag{8.3.13}$$

is the entropy function for the special processes considered. To determine it, one needs measurements of the specific heat at zero stress, the analogue of the $C_0(\theta)$ mentioned in Section 2.5. This will determine $\overline{S}_1$ to within an unimportant additive constant.

Now, we can introduce a kind of analogue of the strain energy function introduced in Section 2.5. Bear in mind that M_1 is considered as a fixed constant, so the empirical functions could depend on M_1 were we to consider other values. Here, we introduce a function

$$W = W(V_1, m, \theta), \tag{8.3.14}$$

such that

$$W(v(\theta), 0, \theta) = 0, \tag{8.3.15}$$

$$\frac{\partial W}{\partial m} = \frac{\partial F_1}{\partial m},$$

$$\frac{\partial W}{\partial V_1} = \frac{\partial F_1}{\partial V_1}, \tag{8.3.16}$$

where the right sides of (8.3.16) are expressed as functions of V_1, m and θ using (8.3.7), (8.3.8) and (8.3.9). The idea is to integrate these equations, subject to (8.3.15), to obtain W. Given suitable isothermal mechanical data, this is enough to determine W uniquely. By arguments similar to those used in Section 2.5, it then follows from above that

$$F_1 = W - \int \overline{S}_1(\theta)\, d\theta. \tag{8.3.17}$$

We will not elaborate on these matters. However, some things are worth noting. There are the constants a and b which must be regarded as rather arbitrary, physically. As is clear from (8.3.7), they do affect the values of chemical potentials, so these are not really so uniquely defined for a given pair of materials. In turn, this affects F_1 and, through it, the dependence of the entropy S_1 on m. For common kinds of stability analyses, such ambiguities seem not to matter. In other kinds of thermodynamic systems, ambiguities of this kind become, in some sense, greater. As a second point, we note that, given sufficiently smooth data, we could calculate $\partial^2 F_1/\partial m \partial V_2$ in two different ways, either by differentiating (8.3.8) with respect to V_1, or by differentiating (8.3.9) with respect to m. Equating the two gives

$$\frac{\partial}{\partial V_1} \ln\left(\frac{V - V_1}{M - m}\right) = \frac{\partial}{\partial m}\left(\frac{M - m}{V - V_1}\right). \tag{8.3.18}$$

Sometimes, relations of this kind are used as a crutch, to try to get the best estimates of the empirical functions from data which might be too sparse or inaccurate to make a very reliable direct determination. Of course, one cannot exclude the possibility that any theory may be overturned by contradictory experimental results, but with rather old, generally accepted theories such as we are considering, I think workers would be more likely to believe that it is the experiment which is in error.

Some interesting observations are reported by Gent and Tompkins [40]. Briefly, a piece of rubber is placed in a pressure chamber which is then filled with a gas. One then waits for some time to let the system come to equilibrium, or at least very close to this. Then the chamber is vented to permit quick escape of the gas. If the pressure was high enough, the whole block explodes, becoming a collection of small fragments. The explanation they propose is as follows. Inevitably, a piece of rubber contains tiny holes. In the first phase, the gas moves rather slowly into the rubber, hence into the holes. Generally, diffusive motions like this are not very fast. As suggested by the above analysis, the holes then fill with gas, at a pressure comparable to that in the gas external to the sample. When the exterior gas is vented, that in the holes cannot move out so quickly, so one still has the high pressure there. Relatively, this is like imposing a high tension on the solid and, intuitively, this could pull it apart. Using this idea, combined with some rough calculations involving rubber elasticity theory, they get an order of magnitude estimate of the value of the pressure which must be exceeded to produce this kind of failure and it is compatible with the observed values. As far as I know, no one else has proposed a better analysis of the phenomenon. Phenomena somewhat like this are sometimes explained in a very different way. If one ignores the possibility that the gas enters the solid, one could argue that the sudden release of the exterior pressure initiates something similar to a tensile shock or stress wave moving into the solid. Generally, such analyses use linear theory, with some *ad hoc* estimate of how large the tension must be to produce breakage, and, usually, this kind of argument predicts breakage at particular places, producing spalling of plates, etc. It is not inconceivable that one could use some analysis of this general kind to explain the phenomenon. Generally, one can encounter cases where two quite different theories seem to explain the same phenomenon. Then, one tries to decide which is best. Decisions of this kind are made, occasionally, but no simple rule book describes how they are made. However, arguments about such matters can result in the design of interesting experiments in order to test the merits of different proposals.

8.4   Exercises

8.1. For the thermodynamic experiments discussed in Section 8.3, one assumption is inappropriate for a solid like dry ice, which readily transfers some of its mass to air. Revise the analysis to apply such solids. To simplify considerations, assume that air does not transfer any of its mass to the solid. You may introduce one other simplifying assumption which seems physically reasonable to you, based on your experience with such solids, after clearly stating what it is. Discuss how the kinds of measurements needed differ from those discussed before.

8.2. In the three-dimensional theory of linear elasticity, W, the strain energy (per unit volume) function, is assumed to be an homogeneous quadratic function of six measures of strain, and that it is positive definite. For a transversely isotropic material, symmetry considerations reduce this to the form

$$W = C_1(x_1 + x_2)^2 + C_2 x_3^2 + C_3(x_1 + x_2)x_3 + C_4(x_4^2 + x_5^2) \\ + C_5(x_6^2 - x_1 x_2),$$

where the C's are material constants and the x's label the strains. What inequalities must the constants satisfy, for the quadratic to be positive, in the strict sense?

9
Equilibrium of Liquid Crystals and Rods

9.1 Liquid Crystal Energies

Although they are liquids, nematic liquid crystals commonly used in display devices involve equilibrium problems which are more like some encountered in solids. Thus, for example, one finds a chapter on them in a volume on elasticity theory in the Landau–Lifshitz series on theoretical physics [41]. At the same time, consideration of them will involve issues which are different from those encountered before. Here, orientation replaces deformation as an important quantity. For equilibrium situations, forces can be involved, but they are generally not of sufficient interest to induce workers to try to measure them and, more often than not, workers solve problems without considering them explicitly. Also, for the first time, we will need to consider effects associated with electromagnetic fields. Additionally, in the design of devices and in measurement of material moduli, transitions play an important role.

These are liquids which are optically anisotropic when they are at rest, being of what is often called the uniaxial kind, as are some crystals like Iceland Spar. Roughly, such materials have one preferred direction, any direction perpendicular to this being physically indistinguishable from any other. Suffice it to say that one can use optical observations to determine the preferred direction. We represent it by a unit vector $\mathbf{n}$,

$$\mathbf{n} \cdot \mathbf{n} = 1, \tag{9.1.1}$$

commonly called the *director*. Actually, the observations do not distinguish between **n** and $-\mathbf{n}$, so

$$\mathbf{n} \ and \ -\mathbf{n} \ are \ physically \ equivalent. \tag{9.1.2}$$

Commonly, observations indicate that in a sample, **n** is not constant but varies with position. Just how it varies is influenced by the nature and sometimes, prior treatment of other materials brought into contact with it, or by placing it in electric or magnetic fields, among other things. The director vector field then defines a kind or orientation pattern and, in turn, what it is affects light passing through the material. By learning how to analyse and control such phenomena workers have been able to design and improve the now familiar practical devices.

From a thermodynamic point of view, we need a constitutive equation for, say, the Helmholtz energy per unit volume. For most fluids, we would think of this as a function of θ, and ρ the mass density. Here, we are dealing with liquids and, as is usually the case for these, it is reasonable to assume that they can be idealized as incompressible materials. Said differently, we can neglect variations in ρ. For the moment, assume that electromagnetic fields are absent. Then, observations indicate that one can have various orientation patterns induced by contact with other materials, etc. Reasonably, different energies are associated with these. As a first guess, workers are likely to try a local theory, using ideas similar to those discussed in Chapter 2 and theory of this kind has performed well. I will sketch the ideas used but omit some derivations. If we introduce rectangular Cartesian coordinates, **n** has three components. First partial derivatives of these with respect to the three coordinates can then be represented by a 3 by 3 matrix, denoted by $\nabla\mathbf{n}$. The basic assumption is that φ, the Helmholtz free energy per unit volume, is of the form

$$\varphi = \varphi(\mathbf{n}, \nabla\mathbf{n}, \theta). \tag{9.1.3}$$

Physical considerations impose some restrictions on the function. From (9.1.2), we should have

$$\varphi(\mathbf{n}, \nabla\mathbf{n}, \theta) = \varphi(-\mathbf{n}, -\nabla\mathbf{n}, \theta). \tag{9.1.4}$$

Also, there is the notion of objectivity, used a little in Chapter 2. Here, the idea is that merely rotating a liquid crystal should not change φ. Any such rotation can be described by a matrix **R**, satisfying

$$\mathbf{R}^T\mathbf{R} = 1, \qquad \det\mathbf{R} = 1. \tag{9.1.5}$$

The requirement is then that, for any such rotation, we should have

$$\varphi(\mathbf{n}, \nabla\mathbf{n}, \theta) = \varphi(\mathbf{Rn}, \mathbf{R}\nabla\mathbf{n}\mathbf{R}^T, \theta). \tag{9.1.6}$$

Roughly, there are two kinds of liquid crystals covered by the description and some others which are not. Typical display devices used in watches, personal computer screens, etc. employ nematic liquid crystals. For these φ is also invariant under reflections. This gives rise to one more condition,

$$\varphi(\mathbf{n}, \nabla\mathbf{n}, \theta) = \varphi(-\mathbf{n}, \nabla\mathbf{n}, \theta), \tag{9.1.7}$$

obtained by using (9.1.6), with $\mathbf{R}$ describing a central inversion ($\mathbf{R} = -1$). The second kind, cholesteric liquid crystals, are used in different sorts of devices, for example those serving as thermometers, by changing color as the temperature changes. For these, φ is not invariant under reflections. Typically, liquid crystal molecules are rather rigid, with one direction large compared to others. In this sense they are rather like the bars we have discussed. Depending on the material, these molecules may or may not display some symmetry with respect to reflections and, usually, it is this which decides whether they are to be regarded as nematics or cholesterics. Here, we will only consider nematics, so (9.1.7) applies.

Another kind of simplification is based on observations. They suggest that, left to themselves, nematic liquid crystals would prefer to have uniform orientation,[1] $\mathbf{n} = \text{const}$. This suggests that such configurations minimize φ

$$\varphi(\mathbf{n}, \nabla\mathbf{n}, \theta) \geq \varphi(\mathbf{n}, 0, \theta). \tag{9.1.8}$$

Granted this, we have, as an analogue of the strain energy functions used before,

$$W(\mathbf{n}, \nabla\mathbf{n}, \theta) = \varphi(\mathbf{n}, \nabla\mathbf{n}, \theta) - \varphi(\mathbf{n}, 0, \theta) \geq 0, \tag{9.1.9}$$

with the property that

$$W(\mathbf{n}, 0, \theta) = 0. \tag{9.1.10}$$

For cases where $\mathbf{n}$ does not vary too rapidly with position, it is reasonable to assume that W can be approximated sufficiently well by its expansion up to terms quadratic in $\nabla\mathbf{n}$ about $\nabla\mathbf{n} = 0$. Imposing all the conditions noted above, one can use rather elementary arguments, discussed by Frank [42], to determine the possible forms of W. This gives, with notation now used by many workers,

$$2W = K_1(\nabla \cdot \mathbf{n})^2 + K_2(\mathbf{n} \cdot \text{curl } \mathbf{n})^2 + K_3\|\mathbf{n} \times \text{curl } \mathbf{n}\|^2$$
$$+ (K_2 + K_4)[\text{tr}(\nabla\mathbf{n})^2 - (\nabla \cdot \mathbf{n})^2], \tag{9.1.11}$$

where the K's are material moduli, functions of θ depending on the material. To have the inequality indicated in (9.1.9) satisfied, these should satisfy

$$K_1 \geq 0, \qquad K_2 \geq |K_4|, \qquad K_3 \geq 0, \qquad 2K_1 \geq K_2 + K_4. \tag{9.1.12}$$

---

[1]In this respect, the cholesterics are different, preferring a kind of twisted configuration, like some to be described later.

This is the so-called Oseen–Zocher–Frank theory, an old workhorse which has satisfactorily described numerous observations and has been part of the basis for designing experiments and devices, etc. It provides a description of what is reasonably regarded as mechanical energy. Also needed is some way of accounting for energies associated with electromagnetic fields and crude estimates do well enough for our purposes. One way to think of it is to consider the liquid crystal as a thermodynamic system, acted upon by fields which are produced externally. The problem is then to estimate the power they supply to the liquid crystal or, what is more pertinent for equilibrium theory, the work they do on the liquid crystal. This is subtle, so it seems worthwhile to consider a simple model to help motivate what is done.

Suppose the liquid crystal is placed on a constant electric field. The force exerted by it on any charge is then obtained by multiplying the (constant) electric vector $\mathbf{E}$ by the charge. We are dealing with dielectrics, their electric response being like the insulating coatings applied to conducting wires, for example. This means that charges able to move very freely are absent. A molecule will, of course, contain the charged electrons and protons, but they remain bound to it, the net charge for the molecule being zero. Thus, the resultant force on the molecule exerted by the constant field is zero. One may conclude from this that the field does no work on the liquid crystal, but this is wrong. Consider the simple case of a dipole, consisting of two point charges, a positive charge q with position vector $\mathbf{r}^+$, a charge $-q$ at $\mathbf{r}^-$. By convention, we associate with them a polarization vector $\mathbf{p}$, given by

$$\mathbf{p} = q(\mathbf{r}^+ - \mathbf{r}^-). \tag{9.1.13}$$

Generally, they will exert forces on each other and be subject to forces exerted by their neighbors, but our concern is more with the influence of $\mathbf{E}$. If, in its presence, the position vectors change a little to $\mathbf{r}^+ + \Delta\mathbf{r}^+$ and $\mathbf{r}^- + \Delta\mathbf{r}^-$, and if ΔW denotes the work done by $\mathbf{E}$, we then have

$$\Delta W = q\mathbf{E} \cdot \Delta\mathbf{r}^+ + (-q)\mathbf{E} \cdot \Delta\mathbf{r}^- = \mathbf{E} \cdot \Delta\mathbf{p}, \tag{9.1.14}$$

which is generally nonzero. Here, the force averages to zero but it still does some work.

In, say, the biaxial stretch of rubber sheets, it is obvious, physically, that the forces are not uniformly distributed along the edges. The work done by them is not likely to be the same as it would be if they were uniformly distributed, when the resultant forces match. Similar remarks apply to loading the bars, etc. Were there some accepted way of correcting for such errors, we would have commented on it in discussing such problems. Here, we do at least have an example of one correction of this kind which is in common usage. Commonly, liquid crystal molecules **do** have a dipole moment, a polarization vector of which one can think as attached to them. **If** they were all aligned the same way, we could simply multiply (9.1.14) by the number per unit volume to estimate the work done per unit volume. However, they are far from being so aligned. One can represent the net effect macro-scopically by a vector $\mathbf{P}$, a kind of statistical average of the individuals in a unit volume. In the absence of an external field, the molecules tend to be randomly aligned, which gives $\mathbf{P} = 0$. Applying a field changes this a little, causing some partial alignment, making $\mathbf{P} \neq 0$. For liquid crystals, the effect is described sufficiently well by a linear equation of the form,

$$\mathbf{P} = a\mathbf{E} + b(\mathbf{E} \cdot \mathbf{n})\mathbf{n}, \tag{9.1.15}$$

where a and b are functions of θ, invertible to give

$$\mathbf{E} = c\mathbf{P} + d(\mathbf{P} \cdot \mathbf{n})\mathbf{n}, \tag{9.1.16}$$

with

$$c = 1/a, \qquad d = -b/[(a + b)a]. \tag{9.1.17}$$

For any fixed value of $\mathbf{n}$, one can use the obvious analogue of (9.1.13), which can be integrated to get W_E, the electric field energy per unit volume.

$$W_E = \frac{1}{2}\mathbf{E} \cdot \mathbf{P} = \frac{1}{2}[c\mathbf{P} \cdot \mathbf{P} + d(\mathbf{P} \cdot \mathbf{n})^2]$$

$$\mathbf{E} = \frac{\partial W_E}{\partial \mathbf{P}}. \tag{9.1.18}$$

This is then a kind of potential energy, describing the work done on the liquid crystal by the field in changing $\mathbf{P}$ from zero to the value of interest. Physically, we expect this to be positive, conditions for this being that

$$c > 0, c + d > 0 \Leftrightarrow a > 0, a + b > 0. \tag{9.1.19}$$

What is neglected here is the induced field. Briefly, a dipole itself produces a field which depends on its polarization. When they are randomly aligned, these fields cancel each other, but there is some net effect when they begin to be aligned by an external field, producing an addition to $\mathbf{E}$, an induced field. This is a relatively small effect, often, but not always, neglected by workers: we will ignore it. In our argument, we assumed $\mathbf{E}$ is constant for simplicity. However, the end result also applies to cases where it varies with position. It should satisfy the equations for a static field, in vacuum, or, more realistically, in air, which are

$$\nabla \cdot \mathbf{E} = \text{curl } \mathbf{E} = 0. \tag{9.1.20}$$

A very similar estimate covers the work done by magnetic fields. In place of $\mathbf{E}$ we have the vector $\mathbf{H}$ describing the imposed magnetic field. It satisfies (9.1.20), with $\mathbf{E}$ replaced by $\mathbf{H}$. In liquid crystals it produces an analogue to $\mathbf{P}$, magnetization described by a vector $\mathbf{M}$. The assumption is that $\mathbf{M}$

and **H** are linearly related, by equations entirely similar to (9.1.15) and (9.1.16). The field energy is then the analogue of (9.1.18),

$$W_H = 1/2\,\mathbf{M} \cdot \mathbf{H}, \qquad (9.1.21)$$

and, again, this should be positive. Again, this neglects an induced field, an effect so small that, as far as I know, no one tries to correct for it.

Often, workers replace W_E by a different expression

$$\overline{W}_E = \frac{1}{8\pi}\,\mathbf{D} \cdot \mathbf{E},$$

where

$$\mathbf{D} = \varepsilon_0 \mathbf{E} + 4\pi\,\mathbf{P}$$

is the so-called electric displacement vector, ε_0 being the dielectric constant for vacuum. One then has

$$\overline{W}_E - W_E = \frac{1}{8\pi}\,\varepsilon_0 \mathbf{E} \cdot \mathbf{E}.$$

This has some merit if one starts to correct for induced fields. Otherwise, one generally considers **E** as a given field, $\overline{W}_E$ and W_E then giving the same energy differences, so it does not matter which of these we use.

Bearing in mind that W_E and W_H represent work done on the liquid crystal, we then subtract these from W to obtain

$$W_T = W - W_E - W_H, \qquad (9.1.22)$$

as the total of the mechanical and field energies per unit volume. For most cases of interest, this covers the energies for which we need to account.

9.2 Orientation by Fields and Walls

First, let us think about what a liquid crystal would prefer to do if it were under the influence of a constant electric field but otherwise were free to do as it wished. Reasonably, it should try to minimize W_T, given by (9.1.22). From our discussion of W, its smallest value is zero, obtained when **n** is any constant vector. As to W_E, we can use (9.1.15) to express this in terms of **E** and **n**, the rest of the energy being

$$-W_E = -\frac{1}{2}\,\mathbf{E} \cdot \mathbf{P} = -\frac{1}{2}[a\mathbf{E} \cdot \mathbf{E} + b(\mathbf{E} \cdot \mathbf{n})^2]. \qquad (9.2.1)$$

The problem is then one of determining the unit vector(s) **n** for which this has the smallest possible value, **E** being given. It makes a difference whether b is positive or negative. The minimizing configurations are easily determined to be of the form

$$b > 0 \Rightarrow \mathbf{n} \parallel \mathbf{E}, \qquad (9.2.2)$$

or

$$b < 0 \Rightarrow \mathbf{n} \perp \mathbf{E}. \qquad (9.2.3)$$

Liquid crystals fitting either case are available. With $b > 0$, one can use $\mathbf{E}$ to induce $\mathbf{n}$ to take a definite direction. This is exploited in modern display devices, for example. Some older devices used materials with $b < 0$, leaving $\mathbf{n}$ free to take any direction in a plane. They did not perform as well and exhibited complex behavior which is hard to analyze.

It is instructive to look at the other equilibria. Taking the differential of (9.2.1), we get the condition

$$-b(\mathbf{E} \cdot \mathbf{n})(\mathbf{E} \cdot d\mathbf{n}) \geq 0, \qquad (9.2.4)$$

for possible $d\mathbf{n}$. These are restricted by the condition that $\mathbf{n}$ be a unit vector,

$$d(\mathbf{n} \cdot \mathbf{n}) = 2\mathbf{n} \cdot d\mathbf{n} = 0, \qquad (9.2.5)$$

being otherwise arbitrary. It is not very hard to show that, for (9.2.4) to hold for all $d\mathbf{n}$ satisfying (9.2.5), $\mathbf{n}$ must be either parallel or perpendicular to $\mathbf{E}$. So, if $b > 0$, one has the stable equilibrium configuration with $\mathbf{n} \parallel \mathbf{E}$ and unstable equilibrium with $\mathbf{n} \perp \mathbf{E}$. Other influences can stabilize the latter and this is used in various ways to produce interesting and useful phenomena. We will return to this later.

There is a different, more mechanical way of understanding these equilibria. By calculating the resultant moment on a dipole, one can motivate the idea that $\mathbf{E}$ produces a couple $\mathbf{L}_E$ per unit volume, given by

$$\mathbf{L}_E = \mathbf{P} \times \mathbf{E} = [a\mathbf{E} + b(\mathbf{E} \cdot \mathbf{n})\mathbf{n}] \times \mathbf{E} = b(\mathbf{E} \cdot \mathbf{n})\mathbf{n} \times \mathbf{E}, \qquad (9.2.6)$$

so the above equilibrium conditions coincide with the condition that $\mathbf{L}_E = 0$. As was suggested before, the constant field gives a zero body force so, with $\mathbf{L}_E = 0$, we have taken care of conditions which may be inferred from ideas of mechanical equilibrium.

Again, the corresponding situation for magnetic fields copies this, replacing $\mathbf{E}$ by $\mathbf{H}$, etc., giving the body couple

$$\mathbf{L}_H = \mathbf{M} \times \mathbf{H}. \qquad (9.2.7)$$

In mechanical terms, the most important effect of the fields is to produce these couples. Roughly, they serve as wrenches which we can use to adjust the orientation, represented by $\mathbf{n}$.

At an interface where a liquid crystal contacts another material, there is a curious effect, long known, but still not very well understood from the view of basic science. Often, the director assumes some definite direction, this being called *strong anchoring*. If not, the experience is that it will make a constant angle with the normal vector to the interface, being free to swing around, subject to this condition, producing what are called *conical boundary conditions.* In detail, what happens depends on the two materials, and, often, on the prior treatment of the other material. For at least some of the more commonly used liquid crystals in contact with an untreated glass plate, **n** is tangent to the plate, favoring no particular tangent direction. One could easily induce it to choose one, by using a magnetic field acting tangentially, for example. However, it would still remain tangent if one applied the field in another direction, at least if the field is not excessively strong. However, it will influence the direction assumed in the interior. Now, consider the glass to be treated, by rubbing it in one direction a number of times. Then **n** will align tangentially,[2] in the direction of rubbing. So, this converts it from a conical case to strong anchoring. Instead of rubbing, one can wash the plate with certain acids or detergents, or use certain coatings, to get **n** to be normal to the plate, another kind of strong anchoring. Other tricks are known for getting strong anchoring, with **n** making other angles with the interface. Suffice it to say that there is a well-developed art of making **n** satisfy one of a variety of boundary conditions at such interfaces. This covers the general ideas commonly used in designing devices and experiments.

9.3   Measurement of Moduli

That W has the form given by (9.1.11) was arrived at by using theoretical ideas to mathematically formulated hunches based on rough observation and not really by curve-fitting any quantitative measurement. Certain kinds of experiments for measuring K_1, K_2 and K_3 have long been used and are not too difficult to analyze, at least roughly. They employ very similar ideas with only slight differences in the set-ups, so we will consider only one in detail, the measurement of K_2. This, like the others, involves a liquid crystal sandwiched between fixed parallel plates. In this case, these samples are prepared to make **n** take up a fixed tangential direction, the same on both plates; one could apply the rubbing technique for glass plates, discussed before. The arrangement is such that the distance between the plates is very small compared to the lateral dimensions of the sample, so it should not matter much what goes on at the edges of the sample. The aim is to get **n** = const., matching the boundary conditions on the plates, and the experience is that one can get a good approximation to this. We will consider only the liquid crystals for which (9.2.2) applies, so an electric field will try to induce **n** to align parallel to it. Actually, workers often use magnetic fields which do have such an aligning effect. As should be clear from the previous discussion, the theory of these two cases

---

[2]One expert informs me that it is not quite tangential, but tilted off by a very small angle.

is essentially the same. To measure K_2, the idea is to apply a constant field, with **E** perpendicular to the constant **n** described above and in a direction tangent to the plates. Then, in this configuration, the body couple given by (9.2.6) is zero, so this is an equilibrium configuration. Although it does not satisfy (9.2.2), the wall has some stabilizing effect on it. To analyze this, we introduce rectangular Cartesian coordinates, with the positive x-axis in the direction of the constant vector **n** described above, the positive y-axis in the direction of the applied field and the z-axis then being normal to the plates.

Think of the region occupied by the liquid crystal as extending to infinity in the x and y directions and in the interval $0 \leq z \leq L$, in the third direction. For the configurations considered, the assumption is that they vary only with z, **n** remaining parallel to the x–y plane. Bearing in mind that **n** must be a unit vector, we can represent it in terms of the angle Ψ it makes with the x-axis,

$$\mathbf{n} = (\cos \Psi, \sin \Psi, 0), \qquad \Psi = \Psi(z). \tag{9.3.1}$$

We assume $\Psi(z)$ is a smooth function which is consistent with the observations. The strong anchoring conditions at the plates give the boundary conditions

$$\Psi(0) = \Psi(L) = 0. \tag{9.3.2}$$

As usual, we assume $\theta = \text{const}$. With (9.1.11), a calculation gives, for the mechanical energy,

$$2W = K_2(\Psi')^2. \tag{9.3.3}$$

With our choice of coordinates,

$$\mathbf{E} = e(0, 1, 0), \tag{9.3.4}$$

e being the magnitude of the field. Then, (9.2.1) gives

$$2W_E = e^2[a + b\sin^2 \Psi] \\ = be^2 \sin^2 \Psi + \text{const.} \tag{9.3.5}$$

For the infinite sample, the total energy will be infinite, so we look at the energy per unit plate area,

$$F = \int_0^L (W \quad W_E) \, dz \tag{9.3.6}$$

Dropping the unimportant constant in (9.3.5), we get

$$2\mathcal{E} = \int_0^L f(\Psi, \Psi') \, dz, \tag{9.3.7}$$

with

$$f(\Psi, \Psi') = K_2(\Psi')^2 - be^2 \sin^2 \Psi. \tag{9.3.8}$$

Of course, any variations of Ψ considered should conform to (9.3.2), and we will want to consider these in much that same way as we did in deriving equilibrium equations for bars, etc. A trick is commonly used in variational equations of this general kind so, for the moment, we ignore (9.3.8). To get equilibrium equations, we proceed, as we have done before to calculate $\delta\mathcal{E}$, getting

$$2\delta\mathcal{E} = \int_0^L \left(\frac{\partial f}{\partial \Psi} \delta\Psi + \frac{\partial f}{\partial \Psi'} \delta\Psi' \right) dz$$

$$= \int_0^L \left[\frac{\partial f}{\partial \Psi} - \left(\frac{\partial f}{\partial \Psi'} \right)' \right] \delta\Psi \, dz + \frac{\partial f}{\partial \Psi'} \delta\Psi \Big|_0^L .$$

(9.3.9)

From arguments such as those used before, the equilibrium equations obtained by setting $\delta\mathcal{E} = 0$ are then

$$\frac{\partial f}{\partial \Psi} - \left(\frac{\partial f}{\partial \Psi'} \right)' = 0. \tag{9.3.10}$$

Here, the last term vanishes because $\delta\Psi = 0$ at the end points, but the trick works for other kinds of problems. Now, by calculation,

$$\left(f - \Psi' \frac{\partial f}{\partial \Psi'} \right)' = \frac{\partial f}{\partial \Psi} \Psi' + \frac{\partial f}{\partial \Psi'} \Psi'' - \Psi'' \frac{\partial f}{\partial \Psi'} - \Psi' \left(\frac{\partial f}{\partial \Psi'} \right)'$$

$$= \Psi' \left[\frac{\partial f}{\partial \Psi} - \left(\frac{\partial f}{\partial \Psi'} \right)' \right].$$

Thus, if (9.3.10) holds, we get a first integral

$$f - \Psi' \frac{\partial f}{\partial \Psi'} = \text{const.} \tag{9.3.11}$$

Conversely, if (9.3.11) holds, (9.3.10) will be satisfied, except perhaps when Ψ is constant. So, one checks (9.3.10) for the possible constant solutions and uses (9.3.11) for the less trivial possibilities. Here, (9.3.10) gives, with (9.3.8)

$$K_2 \Psi'' + be^2 \sin \Psi \cos \Psi = 0 \tag{9.3.12}$$

and there is the constant solution $\Psi = 0$, the only one satisfying the boundary conditions. We anticipated it from the consideration involving L_E. For possible nontrivial solutions, we can use (9.3.11) to replace (9.3.12) by

$$K_2 \Psi'^2 + be^2 \sin^2 \Psi = \text{const.} \tag{9.3.13}$$

Before considering these, let us consider the stability of $\Psi = 0$. For this, it is convenient to use Poincaré's inequality, satisfied by any smooth function Ψ which vanishes at the ends of the interval,

$$\int_0^L \Psi'^2 \, dz \geq \frac{\pi^2}{L^2} \int_0^L \Psi^2 \, dz. \tag{9.3.14}$$

The equality holds when Ψ is a constant times $\sin(\pi z/L)$.

To see where this comes from, one can proceed as follows. Consider the ratio

$$\mathcal{R} = \frac{\int_0^L (\Psi')^2 \, dz}{\int_0^L \Psi^2 \, dz},$$

for smooth nonzero functions Ψ, vanishing at the end points of the interval. This is sometimes called a Rayleigh quotient. Now try to find the smooth function(s) which minimize it.[3] As we have done before with such integrals, we can find extremals by considering $\overline{\Psi}$ as one, writing $\Psi = \overline{\Psi} + \mu\delta\Psi$, where μ is a small parameter, etc. This gives

$$\delta\mathcal{R} = \frac{\delta \int_0^L (\Psi')^2 \, dx}{\int_0^L \Psi^2 \, dx} - \frac{\int_0^L (\Psi')^2 \, dx}{\left(\int_0^L \Psi^2 \, dx\right)^2} \delta \int_0^L \Psi^2 \, dx,$$

which should vanish for $\Psi = \overline{\Psi}$. Let $\overline{\mathcal{R}}$ be the value of $\mathcal{R}$ for $\overline{\Psi}$ and the condition reduces to

$$\delta \int_0^L \left[(\Psi')^2 - \overline{\mathcal{R}}\Psi^2\right] dx = 0,$$

from which it follows that $\overline{\Psi}$ should satisfy the equation

$$\overline{\Psi}'' + \overline{\mathcal{R}}\overline{\Psi} = 0.$$

A general integral of this is a linear combination of $\cos(\overline{\mathcal{R}})^{1/2}z$ and $\sin(\overline{\mathcal{R}})^{1/2}z$. Impose the condition that such a function vanishes at the end points and you find that the coefficients of $\cos(\overline{\mathcal{R}})^{1/2}z$ must vanish and that

$$\overline{\mathcal{R}} = n^2\pi^2/L^2, \qquad n = \pm 1, \pm 2, \pm 3, \dots .$$

Clearly, the smallest such value occurs for $n = 1$, and this gives (9.3.14).

Also, it is easy to verify that

$$\sin^2 \Psi \leq \Psi^2. \tag{9.3.15}$$

Thus, from (9.3.7) and (9.3.8),

$$2\mathcal{E} = \int_0^L [K_2(\Psi')^2 - be^2 \sin^2 \Psi] \, dz$$

$$\geq \left(K_2 \frac{\pi^2}{L^2} - be^2\right) \int_0^L \Psi^2 \, dz. \tag{9.3.16}$$

---

[3] For a complete and rigorous treatment, one needs to show that the minimum is attained by some such function. A proof of this is given on p. 122 of Courant and Hilbert [43]. One can use simple functions, for example, the trigonometric functions mentioned below, to show that $\mathcal{R}$ can be arbitrarily large, so $\mathcal{R}$ does not take on a maximum value for any smooth function.

Obviously $\mathcal{E} = 0$ when $\Psi = 0$ so $\Psi = 0$ will minimize $\mathcal{E}$ whenever the last expression is nonnegative. That is

$$be^2L^2 \leq K_2\pi^2 \Rightarrow \Psi = 0 \text{ is stable.} \tag{9.3.17}$$

To explore this when the inequality fails, we can use the second derivative test. For this, we proceed as usual. This amounts to taking $\Psi = \mu\delta\Psi$, where μ is a small parameter, and approximating $\mathcal{E}$ to quadratic terms in μ. This gives

$$2\mathcal{E} \cong \mu^2\delta^2\mathcal{E} = \mu^2 \int_0^L [K_2(\delta\Psi')^2 - be^2(\delta\Psi)^2]\,dz$$

$$\geq \mu^2 \left(K_2\frac{\pi^2}{L^2} - be^2 \right) \int_0^L (\delta\Psi)^2\,dz, \tag{9.3.18}$$

with (9.3.14) being used again, now applied to $\delta\Psi$. Also, the equality will hold for $\delta\Psi = \sin(\pi z/L)$. If $\delta^2\mathcal{E} < 0$ for any admissible choice of $\delta\Psi$, $\Psi = 0$ is unstable so

$$be^2L^2 > K_2\pi^2 \Rightarrow \Psi = 0 \text{ is unstable} \tag{9.3.19}$$

and the liquid crystal must then adopt some other configuration to be in stable or metastable equilibrium. The borderline defined by the change in stability of $\Psi = 0$ gives a critical value e_c of the field strength e, given by

$$be_c^2L^2 = K_2\pi^2. \tag{9.3.20}$$

The idea of the experiment is to determine e_c. That is, one adjusts e to find the value at which the uniformly oriented configuration $\Psi = 0$ shifts to another kind. One can use optical methods to detect this, although workers have found other methods to be more accurate. Of course, L can be measured. With (9.3.20) this gives an experimental estimate of K_2/b. Another kind of experiment is used to determine b, but we will not discuss it. With the two, one also gets an estimate of K_2 at the prevailing temperature. This modulus has force as its physical dimension. Of course, it varies with the material and, for a given material, with temperature. Typically, it is of the order of 10^{-6} dynes, which is an exceedingly small force by normal standards.

Roughly, it is this which lets the devices operate with a very small supply of power. It is worth noting that, according to elementary electromagnetic theory, $V = eL$ has the physical dimensions of the electrostatic potential, commonly meausured in volts, so (9.3.20) could be written in terms of a critical "voltage" V_c

$$bV_c^2 = K_2\pi^2. \tag{9.3.21}$$

In this arrangement V_c has no very clear physical significance. Devices and other experiments commonly employ fields directed normal to such plates and involve similar bifurcations. Then the analogous critical voltage space is the voltage across the plates. Typically it is a few volts. In this form it is independent of L, a prediction which agrees with experience.

One can proceed to analyze the nontrivial equilibria. It turns out that as one increases e, one passes through critical values, marking places where the number of solutions of the equilibrium equations jumps to a higher value. One can analyze these to determine which are the energy minimizers, but I will not try to discuss this. Actually, the minimizer has the properties one might expect intuitively. As soon as e gets larger than e_c, one gets a compromise between the wall and field effects. At the wall, the wall has its way, maintaining the boundary conditions. Away from the wall, the director rotates a little, a small step towards becoming parallel to the field. As one moves away from the walls it rotates more, reaching a maximum at the mid-plane $z = L/2$. It is easy to see that if $\Psi(z)$ is a solution, so is $-\Psi(z)$, or, said differently, it could rotate either to the left or the right, a prediction not contradicted by experience. The behavior is somewhat similar to that which we have encountered before, in the "pitchfork" bifurcations, and some workers use the name for them.

For one of the two solutions we will have Ψ increasing with z, from zero at $z = 0$, to some maximum value $\Psi_0 < \pi/2$ at $z = L/2$. Since it is a maximum, we will have

$$\Psi_0 = \Psi(L/2), \qquad \Psi'(L/2) = 0. \tag{9.3.22}$$

Then, (9.3.13) takes the form

$$K_2(\Psi')^2 + be^2 \sin^2 \Psi = be^2 \sin^2 \Psi_0, \tag{9.3.23}$$

from which it is clear that

$$\sin^2 \Psi \leq \sin^2 \Psi_0. \tag{9.3.24}$$

We now do some juggling. Let

$$k = e\sqrt{b/K_2}. \tag{9.3.25}$$

It is a number with the reciprocal of length as its physical dimension. With (9.3.24), we can introduce an angle χ satisfying

$$\sin \chi = \sin \Psi / \sin \Psi_0, \tag{9.3.26}$$

varying from zero to $\pi/2$ as Ψ varies from zero to Ψ_0. In terms of these quantities, a routine calculation gives, as a replacement of (9.3.23),

$$(\chi')^2 = k^2[1 - \sin^2 \Psi_0 \sin^2 \chi], \tag{9.3.27}$$

the conditions on χ being

$$\chi(0) = \chi(L) = 0, \qquad \chi\left(\frac{L}{2}\right) = \frac{\pi}{2}. \tag{9.3.28}$$

Consider the solution for $0 \leq z \leq L/2$ where $\chi' > 0$. Then (9.3.27) and (9.3.28) become

$$kz = \int_0^\chi \frac{d\varphi}{\sqrt{1 - \sin^2 \Psi_0 \sin^2 \varphi}} \tag{9.3.29}$$

and

$$\frac{kL}{2} = \int_0^{\pi/2} \frac{d\varphi}{\sqrt{1 - \sin^2 \Psi_0 \sin^2 \varphi}} = h(\Psi_0). \tag{9.3.30}$$

Given k and L, the idea is to solve (9.3.30) for Ψ_0. Substituting this in (9.3.29) gives z as a function of χ, the inverse of the function $\chi(z)$ for $0 \leq z \leq L/2$. To obtain the solution in the upper half, put

$$\chi(z) = \chi(L - z), \qquad L/2 \leq z \leq L, \tag{9.3.31}$$

it being a simple matter to verify that this gives a solution of (9.3.27) satisfying (9.3.28).

Concerning $h(\Psi_0)$, given by (9.3.30) for $0 \leq \Psi_0 < \pi/2$, one can show that

(a) $h(\Psi_0)$ is a monotonically increasing, continuous function of Ψ_0;

(b) $\lim_{\Psi_0 \to \pi/2} h(\Psi_0) = \infty$;

(c) $h(0) = \frac{\pi}{2}$.

This means that kL determines a unique value of Ψ_0 in the range of interest provided

$$kL = eL\sqrt{\frac{b}{K_2}} > \pi. \tag{9.3.32}$$

From (9.3.19), this means that these solutions occur whenever conditions are such that $\Psi = 0$ is unstable. For these nontrivial solutions, it follows from (a) and (c) that as kL approaches the critical condition indicated by (9.3.20), Ψ_0 approaches zero, so near this, it and hence $\sin^2 \Psi_0$, will be small. Then (9.3.13) implies that Ψ will be small, close to zero. So, the distortion develops rather smoothly as one increases e, passing through e_c, again like the pitchfork bifurcations we studied before. Actually, this makes it a little difficult to observe exactly when the transition occurs: other methods can detect it, before it is optically discernible.

As one increases e, or, if you like, the somewhat fictional voltage eL, kL increases so, from (a), Ψ_0 increases, $\mathbf{n}$ becoming more nearly parallel to $\mathbf{E}$ near the mid-plane. The experience is that it only takes a few volts to get $\mathbf{n}$ fairly close to the direction of $\mathbf{E}$ in a large fraction of the gap. In the small parts near the walls, $\mathbf{n}$ then varies rapidly with position to adjust to the wall orientation.

Transitions somewhat similar to this occur in various other situations of physical interest. To recall what we did, we recognized that, with an electric (or magnetic) field, we could have unstable equilibria, with **n** perpendicular to the field. We then introduced something else, the wall effects, to help stabilize this. As we have seen, this can be successful if the field is not too strong. If it is strong enough it gives a transition. Workers refer to the transitions of this kind as *Fréedericsz transitions* in honor of a Russian scientist who discovered them experimentally using magnetic fields.

The classical methods for measuring K_1 and K_3 employ rather similar set-ups and transitions. For K_1, we start with the sample oriented as before, but apply a field in the direction normal to the plates, one case where the voltage across the plates is the product of the field strength and the gap width. For K_3, the plates are treated to make **n** align normal to them. The field is then applied perpendicular to this direction. The corresponding calculations are discussed by various writers. For example, they are included in three of the best general reference works on liquid crystals, the books by Chandrasekhar [44], de Gennes [45], and Virga [46]. Display devices exploit Fréedericsz transitions, using fields strong enough to induce **n** to become nearly parallel to the field except very near the walls. Chandrasekhar discusses the twisted nematic cell, which is used in many such devices. The mathematical theory of nematic liquid crystals has undergone a period of rapid development, as is discussed by Virga [46].

There is a point worth bearing in mind. We considered only some very special types of configurations. Those regarded as stable may turn out to be unstable with respect to more general variations. Cohen [47] explored this theoretically for the set-up used to measure K_1. He finds that, for many values of the moduli, there is an instability which sets in at a field strength lower than that given by one-dimensional theory, causing **n** to be nonuniform in a periodic manner. Conversation with experimentalists indicates that some had noticed a little undulation which could be associated with this, not so small in recently discovered liquid crystals where K_1 is relatively large, so there is some reason to be concerned about this. So far at least, it has not been feasible to analyze the sample of finite size which must be used in the experiments. So, as is typical, one gets into more sophisticated problems as one tries to take care of the inevitable loose ends, in order to understand nature a little better.

9.4 Elastica Theory

It is a matter of common experience that the straight long thin bodies we call rods, bars, beams and wires tend to depart from their straight shape, to buckle when we apply compressive loads of moderate size at their ends. Obviously, the bar theory discussed earlier cannot cope with this. One can use somewhat similar one-dimensional theory, but it must let the line segments become curves. The simplest possibility is that the curve is contained in some plane, and, for a number of physically interesting situations, this is the case. The simplest theory of this kind is Euler's theory of the Elastica, created by this famous scientist in the eighteenth century. It has proved to be very good, so it is still used by those concerned with designing safe structures. It may seem strange to group these rods with liquid crystals. The reason is that, mathematically, the more elementary analyses of Fréederícsz transitions are almost the same as the Euler buckling problems. For this reason, functions like the $h(\Psi_0)$ occurring in (9.3.30), one of the so-called elliptic integrals is, much like the trigonometric functions, long considered to be important enough to have their values listed in tables of functions, now readily calculated by computer.

The notion of a reference configuration can be carried over unchanged from bars: the stable unstressed configurations are considered as straight, either way. However, we prefer to use different notation. Instead of x, we will write S, with

$$0 \le S \le \ell \qquad (9.4.1)$$

describing the reference. A deformation will take this to some plane curve, so we introduce rectangular Cartesian coordinates (x, y) in the plane. The plane curve can then be represented parametrically by equations of the form

$$x = x(S), \qquad y = y(S). \qquad (9.4.2)$$

We can also introduce an analogue of the $y(x)$ used in rod theory; it is the arc length s, the distance measured along the curve, given by[4]

$$s = \int_0^S \sqrt{(x')^2 + (y')^2}\, dS = s(S). \qquad (9.4.3)$$

In terms of this, we can introduce the stretch

$$\lambda = s', \qquad (9.4.4)$$

describing changes in length in the same way it did in bar theory. In buckling phenomena, the bending deformations are generally large enough to be perceived by the eye, but the changes of length tend to be relatively small. The classical theory introduces an approximation associated with this which does simplify the theory. It is that the rod is considered to be inextensible, incapable of changing its length, so, always,

$$s = S, \qquad \lambda = 1 \Rightarrow x'^2 + y'^2 = 1. \qquad (9.4.5)$$

This makes the vector (x', y'), tangent to the curve, a unit vector, representable in the form

$$x' = \cos \nu, \qquad y' = \sin \nu, \qquad (9.4.6)$$

---

[4]Readers not familiar with this can find a discussion of it, curvature, etc., in any elementary book on differential geometry.

ν being the angle it makes with the x-axis. To cover the fact that a rod offers some resistance to bending, a bending energy per unit length W is introduced, a kind of strain energy function. For the ancient theory, the assumption is that it is of the form

$$W = W(\theta, \nu') = \frac{1}{2} K(\nu')^2, \tag{9.4.7}$$

where K is a positive function of θ called the *flexural rigidity*. From differential geometry ν' is the curvature of the curve. Different kinds of boundary conditions, etc., are used to model different kinds of physical problems. Often, but not always, they are such that the thermodynamic ideas of equilibrium are applicable. We will consider one case where they are, as an illustrative example. Suppose a rod is set into concrete, forming part of a floor, with the rod normal to the floor. Choose axes so that the floor is the plane $x = 0$, the rod's reference configuration being on the x-axis. We will regard one end of the rod as being at the origin, ignoring the piece buried in the concrete except for estimating the boundary conditions. Physically, setting it in will fix the end position and also fix the tangent vector at this end, say $S = 0$, giving the boundary conditions

$$x(0) = y(0) = \nu(0) = 0. \tag{9.4.8}$$

Physically, no work will be done on this end. The sides are to be left free so no work will be done here. As the words suggest, we consider the system to be in the earth's gravitational field, giving a gravitational force acting in the direction of the negative x-axis. We ignore the effect of this on the rod itself. However, we will use it to dead-load the end $S = \ell$. That is, we will here firmly attach a rigid body with weight w. Associated with this is a potential energy which can be taken as

$$wx_c, \tag{9.4.9}$$

where x_c is the x-coordinate of the centre of mass of this body. For simplicity, we assume it attached so that this point is also the end of the rod, so

$$x_c = x(\ell). \tag{9.4.10}$$

So, overall, the loading is conservative and, as usual, we assume $\theta = $ const., one of the standard situations covered by equilibrium theory. The applicable thermodynamic potential can be taken as

$$\hat{\mathcal{E}} = \frac{1}{2} \int_0^\ell K(\nu')^2 \, dS + wx(\ell). \tag{9.4.11}$$

Here, there is a possibility of having end-point minima, with the weight and perhaps part of the rod, coming into contact with the floor. Clearly, this depends on the size and shape of the weight. We will ignore this constraint and, for this reason, our analysis will be incomplete.

In part, the boundary conditions (9.4.8) can be taken care of, by using (9.4.6) to write

$$x(S) = \int_0^S \cos \nu \, dS, \qquad y(S) = \int_0^S \sin \nu \, dS. \qquad (9.4.12)$$

With this, we can rewrite (9.4.11) in the form

$$\hat{\mathcal{E}} = \int_0^\ell \left[\frac{1}{2} K(\nu')^2 + w \cos \nu \right] dS, \qquad (9.4.13)$$

with ν still subject to the boundary condition $\nu(0) = 0$. To determine the conditions needed for equilibrium, we can, as usual, calculate $\delta\hat{\mathcal{E}}$ and equate it to zero. In the interior of the interval, we get the equation which one would expect from (9.3.10),

$$Kv'' + w \sin \nu = 0,$$

and the first integral suggested by (9.3.11). In addition, the analogue of the last term in (9.3.9) produces another boundary condition: we should have

$$\nu(0) = \nu'(\ell) = 0. \qquad (9.4.14)$$

Essentially, this is the liquid crystal problem we analysed earlier. To see this, put

$$\nu = 2\Psi, \qquad \ell = L/2, \qquad (9.4.15)$$

so we have

$$\Psi(0) = 0, \qquad \Psi'(L/2) = 0, \qquad (9.4.16)$$

fitting conditions listed in (9.3.2) and (9.3.22). Also, (9.4.13) becomes

$$\hat{\mathcal{E}} = \int_0^{L/2} [2K(\Psi')^2 + w(1 - 2\sin^2 \Psi)] \, dS$$

$$= 2 \int_0^{L/2} [K(\Psi')^2 - w \sin^2 \Psi] \, dS + \text{const.}, \qquad (9.4.17)$$

the integrand being of the same form as in (9.3.16) with the constants labeled in a different way. The upper limits look different so, in effect, we are looking at the liquid crystal energy for the lower half of the sample. However, from symmetry, it is easy to see that the upper half has the same energy, so one can replace the upper limit in (9.3.16) by $L/2$ and multiply the new integral by 2. Thus, with some bookkeeping, one can use the whole analysis for these two problems, which are physically quite different. The critical condition corresponding to (9.3.20) works out to be

$$w_c \ell^2 = K\pi^2. \qquad (9.4.18)$$

Thus, for $w < w_c$ the rod stays as is in its reference configuration. For w a little larger than w_c, it will start to bend over a little. As the load increases, $\nu(\ell)$ will get closer to π, the rod bending nearly double. Realistically, somewhere along the line it will bend enough to let the weight contact the floor. Consideration of this would, of course, make the two problems different.

In this case, it is not customary to try to use (9.4.18) to obtain an experimental value for K, partly because small misalignments etc. introduce noticeable errors in such measurements. A classical estimate which seems good gives $K = EI$, where E is Young's modulus, obtainable from a simple tension test and I is a geometrical factor, a certain moment of inertia. One can get it directly using a simple bending experiment, for example. These and some other aspects of classical rod theory are discussed in more detail by Love [48], for example.

In previous studies we have concentrated more on issues relating to attempts to find equations to describe response of materials. Here, this issue is fairly well settled and we move into the province of the engineer who needs to resolve stability questions in order to design safe structures. Such workers use various ideas which have not been mentioned in our discussions, as can be seen from the books of Leipholz [49] or Thompson and Hunt [25], for example. Numerous other kinds of workers are interested in aspects of stability, producing a literature so large as to be indigestible for one person.

9.5 Exercises

9.1. An unusual stability problem arises in liquid crystals, associated with what is called the "plage tordue." The sample is prepared like that discussed in Section 9.3 except that no field is applied. Instead, one plate is simply rotated relative to the other, keeping them parallel, so the orientation at the two plates becomes different. To analyze this, use (9.3.3) to calculate the energy. Then determine the equilibrium equation and boundary conditions, and find the solutions. You should get two solutions for each angle of rotation, if you read Section 9.1 carefully. Determine which of these is most stable and how this changes as the angle of rotation increases.

9.2. An experiment long used to determine values of K_1 in (9.1.11) employs a sample of the kind described in Section 9.1, except that the field is here normal to the plates. Instead of (9.3.3), assume that

$$\mathbf{n} = (\cos \chi, \sin \chi), \qquad \chi = \chi(z).$$

Substituting this in (9.3.3) gives

$$2W = (K_1 \cos^2 \chi + K_3 \sin^2 \chi)(\chi')^2, \qquad \chi = \chi(z),$$

an assertion you should check. Calculate the function describing the field energy, then derive the equilibrium equation and boundary conditions. If you have trouble getting this equation, review Section 3.1.

9.3. For the situation just described, determine whether the equilibrium equation has any trivial solutions satisfying the boundary conditions, then find an integral of the equation. How does what it implies about trivial solutions compare with your findings?

9.4. Can you find a physical problem for the Elastica which is essentially the same, mathematically, as that discussed in Exercise 9.3, perhaps for special values of the constants K_1 and K_3?

10

Reconsideration of Generalities

10.1   Systems, Energy, and Temperature

Some general ideas about classical thermodynamics were mentioned briefly in Chapter 1, later chapters providing some examples of how they are used in practice. One needs to obtain a better grasp of them to avoid misuse and to begin to understand some of the real difficulties involved in their application to some phenomena encountered, particularly in solids.

Firstly, not every physical system can be properly regarded as a thermodynamic system fitting the ideas of classical thermodynamics. It is generally agreed that this is a macroscopic theory, not applicable to an electron, for example. However, various kinds of macroscopic systems are also excluded. In particular, the first law should be applicable and it involves two ideas. One is the equation relating changes in energy to power and heating. The other is a relation of energy to states. In this section, we will consider the first. Our study of mixtures suggested one limitation based on consideration of this, that a thermodynamic system should contain a fixed set of matter. Occasionally, workers do consider systems which do not contain fixed sets of matter, but it is somewhat tricky to deal with these in a physically sound manner.

In our study of liquid crystals, we encountered those electric and magnetic fields which do not vanish in parts of space where there is no matter, that is "vacuum fields." More generally, this includes time-dependent electromagnetic fields describing radiation, for example. According to electromagnetic theory there is an energy associated with such fields. For, say, a fixed region of space, the time derivative of this energy is related to the flux of energy through the boundary. However, in general, there is no reasonable way to consider this flux as representing power or heating. If you recall, in the liquid crystal study we did not try to treat the vacuum fields as thermodynamic systems. They did interact with the important part of the thermodynamic system there considered, the liquid crystal. There, we were able to account for the interaction in a way which makes sense in terms of the first law, as it applies to the liquid crystal. From this, we see that sometimes, something which is not properly regarded as a thermodynamic system can be considered to interact with a thermodynamic system. Now, if one tries to construct a general, systematic theory of thermodynamics, it makes life much easier to assume that a thermodynamic system can only interact with other systems of this kind, so the first law, etc., can be applied to one, or any combination of these. So we bent this rule to make possible the analysis of a situation of interest. Appreciate that one is taking a risk in bending or breaking this rule.

The rule has been broken in a different way with good results in the application of thermodynamics to black-body radiation. This involves radiation in a cavity with rigid walls, maintained at constant temperature and also opaque to radiation. This means that, at the walls, the flux of electromagnetic energy is to vanish. In his discussion of this, Pippard writes (in Chapter 6 of [4]) that,

> *It is not at all obvious that we are justified in applying thermodynamics to such a problem, since the formulation of the laws of thermodynamics was based on experience of material bodies, and we need feel no a priori confidence that they are of sufficiently wide validity to embrace mechanical processes in which radiation provides the motive force.*

Read what else he said and you will see that he is not expressing doubts about the end results. As I see it, there is here the implication that something is wrong with the old laws. If the first law really applied, why should it be necessary to make the walls opaque? Actually, with our liquid crystal studies, we would have encountered more serious difficulties had we tried to introduce corrections for induced fields. In various crystals, application of stress induces electric fields. Classical thermodynamics cannot deal satisfactorily with such phenomena and, in particular, one needs to consider modifications of the first law. Relevant theory of this kind is discussed by Grindlay [50] and Brown [51], for example.

There is some consensus of opinion that the energy equation for bars considered in Chapter 2 remains applicable when the phenomena of plasticity and/or viscoelasticity occur. However, other difficulties arise in trying to apply ideas of classical thermodynamics to such situations. Later, we will elaborate on this and discuss some of the ideas which remain applicable.

In dealing with behavior near phase transitions, workers rather frequently introduce a modification of theory which goes under various names, being called *Cahn–Hilliard theory, Landau–Ginzberg theory, Korteweg theory,* or *Van der Waal's theory.* About a century ago, van der Waals [52] first considered theory like this for fluids, deduced from consideration of molecular theory. Roughly, it involves taking the commonly used formula for the Helmholtz free energy density and adding quadratic terms in higher derivatives. The intuitive idea is that these will be small terms, not affecting predictions much, except when the usual form suffers the loss of convexity associated with spinoidal regimes. Then, the experience is that it does produce some significant and interesting modifications in predictions. For our thermoelastic theory of bars, the usual function $\psi(\lambda, \theta)$ would be replaced by

$$\overline{\psi} = \psi(\lambda, \theta) + a \left(\frac{\partial \lambda}{\partial x} \right)^2, \tag{10.1.1}$$

where a is a positive constant. Internally consistent three-dimensional theory of this kind is discussed by Dunn and Serrin [53]. To obtain this consistency, one needs to modify the format used before. The usual equation of motion (2.3.8) still applies but the constitutive equation for the stress σ changes. The effect is to replace an equation which was of second order in $y(x, t)$ by an equation of fourth order. Mathematically, one then expects to need an additional boundary condition. It is not entirely clear how best to select boundary conditions to model particular physical situations. Some try to avoid dealing with this by considering bars of infinite length. Others make some definite choice which seems reasonable to them, but there is a lack of hard evidence bearing on the correctness of such choices. So, in this respect, this kind of theory is somewhat shaky, although it is an old theory which has generated a sizeable literature. In terms of thermodynamics, a difficulty arises in considering the energy equations discussed in Section 2.3; one finds that it needs to be modified. If one tries to match this to the first law, the expression for power should now take the form

$$P = \left(\sigma \dot{y} + \overline{\sigma} \frac{\partial \dot{y}}{\partial x} \right) \Big|_0^L + \int_0^L f \dot{y} \, dx, \tag{10.1.2}$$

where $\overline{\sigma}$ is given by a certain constitutive equation. In terms of intuitive ideas about power, it is at least difficult to motivate inclusion of the additional term. If we accept it, we have other complications. For example, if we set $f = 0$ and clamp the ends so that $\dot{y}(0, t) = \dot{y}(L, t) = 0$, this is not sufficient to attain mechanical isolation ($P = 0$). Mathematically, it would be quite acceptable to obtain $P = 0$ by adding boundary conditions making $\partial \dot{y}/\partial x = 0$ at the ends. However, physically it is not clear as to what we should do to realize this. So, for example, it is at least tricky to analyse possible equilibria using theories of this kind and the difficulties are associated, in part, with trying to make good sense of the first law. Actually, one needs a modification similar to (10.1.2) for a reasonable thermodynamic treatment of the older theory of the Elastica, mentioned in Chapter 9, but for it, the problem of realizing analogous conditions at the ends are better understood. Somewhat similar quirks occur in other theories of rods, plates and shells.

In liquid crystal theory, power is considered to include the possibility that one can do work by causing the director to change, as well as by causing samples to move, as is discussed in references cited in Chapter 9. One needs to bear this in mind in interpreting what it means for such a system to be mechanically isolated, for example. In later discussion of the laws of classical thermodynamics, remember that workers can and do use such reinterpretations to try to extend their range of applicability. It does seem difficult to justify such practices on an *a priori* basis, but some good theory would not exist if such practices had not been used.

As a general rule, writers of elementary texts on thermodynamics focus on theories fitting a certain pattern, to be discussed more later. In previous chapters, we used theories which do fit this mould fairly comfortable. As is suggested by the remarks made above, in various situations which are commonly encountered, one meets theories which do not fit the mould so well. The better one understands the general ideas, the better are one's chances of adapting these to apply classical thermodynamics to theories which are, in one way or another, awkward.

As a matter of common experience, you know that if you take a mercury thermometer from a warm room and plunge it into cold water, it takes some time before the thermometer reads what you can believe to be the temperature of the water. If, for some reason, the water temperature were changing rapidly with time, such a thermometer could be put into it and read, but the readings would be rather meaningless, physically. A different kind of thermometer may adjust more quickly and provide meaningful measurements of temperature. A situation of this kind does occur in turbulent flows. In air, workers have used hot-wire anemometers to get what are regarded as meaningful measurements of temperature, for example. Said differently, we want the thermometer to adjust quickly enough to be in equilibrium, or very close to it, at each time, although the system may be quite far from equilibrium. Similarly, if the temperature in the system varies rapidly with a position, a large thermometer can at best provide some average of the temperature in the system. So if a smaller thermometer is used, one would expect to get a different reading. If it can be made small enough to provide a reading which we can believe, the thermometer is acceptable, as a practical matter. Again, the demand is that the thermometer be in equilibrium, or at least very close to this, although the system is not. Essentially, classical thermodynamics presumes that it is possible to find thermometers able to provide meaningful measurements: the classical theory of thermometers is an equilibrium theory. One implication is that absolute temperature is here considered to be a meaningful concept in a system undergoing nonequilibrium processes. At least, this is my view of classical thermodynamics. If you converse with thermodynamicists, you may well find some contradictory opinions and for a more basic treatment, one really should start with a more primitive idea of empirical temperature. Certainly, I concede that this classical idea has its limits. My view is that, when one exceeds these, one may well need to think about temperature in quite a different way, if at all.[1] In molecular theory, one encounters a different notion of temperature, related to kinetic energy of molecular motions. It then becomes a tenable notion that different species in a mixture can have different temperatures, at what is, from a macroscopic view, the same point, for example. Classical thermodynamics does have its faults and virtues, but one cannot make one's own judgment about these, if one does not understand the subject. Suffice it to say that there are other, rather different kinds of thermodynamics beyond the scope of this work. Briefly, we will, as before, use absolute temperature freely.

Bear in mind that different workers can have somewhat different views of exactly what should be meant by classical thermodynamics. For this reason, I am trying to make my own view reasonably clear.

10.2   General Processes

For various reasons, one needs some picture of all of the processes which can occur in a thermodynamic system. Ideally, we would like to know everything to be known about them. If we attain this goal for a system, we have mastered the problems which may be involved in trying to apply thermodynamics. Thus, in practice, we will start with a picture which is incomplete and possibly incorrect, to be rectified as our understanding improves.

Realistically, we impose some limits on what can be done to the system to develop manageable theory. For, say, a wooden beam of good quality, we could use our theory of thermoelastic bars for some purposes. Or, if we were concerned with buckling we could try Euler's theory of the Elastica. With the theory of mixtures, we open the door to exploring effects in the beam produced by changes in humidity. Obviously, it would take a much more complicated theory to deal with all these effects and we would still exclude some possibilities such as setting the beam on fire. So, what is to be considered as the set of possible processes in the beam depends on the judgment and concerns of the worker who may be interested in it and these may change as the worker learns more from studies based on the picture adopted at the start. We saw this in the equilibrium theory, when some considerations induced us to abandon the notion that equilibria be free of

---

[1]Obviously, there are conceptual difficulties involved in trying to associate a temperature with those vacuum fields, for example.

discontinuities. In this, we did not explore as to whether our nonequilibrium theory is capable of describing such topics as the evolution of our "phase mixtures" from initially smooth configurations by time-dependent processes and, if so, whether this involves initiation of shock waves, for example. If it cannot cope with such phenomena, this theory is in difficulty. In pondering such possibilities, it is a good idea to bear in mind that a variety of different nonequilibrium theories can become indistinguishable for typical equilibrium analyses. For the case at hand, various versions of viscoelasticity theory would, in this sense, agree with our thermoelasticity theory. So, if a worker encounters some obstacle of the kind indicated, he may well try some other nonequilibrium theory in order to find a way of eliminating the difficulty.

We have just introduced a notion which, in effect, says that we can change the set of possible processes by making some change in the constitutive equations. If you accept this, you may have some qualms about arguments made in Chapter 2 in deducing restrictions on constitutive equations. When we considered those processes, we assumed that those "sources" f and r could be prescribed arbitrarily. One finds some workers proceeding in a way which I interpret as meaning that they tacitly accept assumptions of this kind. Some, for example Coleman and Noll [54], do make such assumptions explicit. This has the effect of eliminating, or at least diminishing, the influence of the form of the constitutive equations on the set of possible processes. There is another school of thought, illustrated by the work of Müller [55], for example, which I do find attractive. According to this, such 'source terms' are to be assigned only in ways which are physically realistic. Of course, this involves some exercise of judgment. For the bar theory, one person might decide that it is only reasonable to have $f = 0$ and either $r = 0$, or that r be given by Newton's law of cooling, for example. This has the effect of making the set of possible processes smaller in what is essentially the same theory. With the latter approach, one has to work harder to get the implied restrictions on constitutive equations and one may or may not obtain the same restrictions for a given kind of theory. What we should do is to rederive the restrictions, using the different assumptions, to see if we do arrive at the same or different conclusions. I will leave it to the interested reader to try this for himself. It may be helpful to examine how Müller does this, using a "Lagrange multiplier" theorem due to Liu [56]. I will not belabor this. However, it is my view that the general scheme should be sufficiently flexible to leave the decision of how best to define the set of possible processes to the individual. Different workers could come to contradictory conclusions about the same system. Then is the time to examine the differences carefully and objectively in an attempt to decide which view is better.

10.3   Static and Reversible Processes

Of course, a process can be independent of time, being then what I will call a *static process*. These are the only processes which we considered in our various equilibrium studies and, at least partly because of this, some thermodynamicists like to call these "equilibrium states." I dislike the name for two reasons. Consider a spacecraft not using its mean of propulsion in outer space, so it can be regarded as being subject to no forces. From commonly used ideas in mechanics, its total linear momentum and moment of momentum should be constant, so it cannot come to rest if these do not vanish. Consider it to be thermally insulated, making it an isolated system. Then, it is perfectly reasonable to think that it will come to equilibrium according to the definition we have used for this, mentioned in Chapter 1. One has to take into account the aforementioned conditions in deciding what are the "possible variations," but one always needs to worry about this. There are also good precedents in the old theories of planets. Some discussion of how examples of this kind fit into the thermodynamic scheme is given by C. S. Man [57]. Older literature on this includes a massive and famous work by Poincaré [58] which he clearly labels as a study of the equilibrium of rotating fluid masses, following still older practice. It is then misleading and somewhat blinding to think that only static processes can qualify as equilibria. As a second point, in my opinion some static processes should not be regarded as equilibria. At least the studies we have made all suggest that, in equilibrium, the temperature should be independent of position. With our rigid bars, we may obtain steady state heat conduction by insulating the sides and keeping the ends at different constant temperature. To me, this is not equilibrium although it is a static process. Rather commonly, thermodynamicists seem to prefer to regard "equilibrium" as a primitive concept, providing some verbal description of what they mean by this. Those I have examined conform to my understanding of a static process, I think. There seems to be a common prejudice that time-dependent reversible processes are not really attainable, but "equilibrium states" are. I do accept, as mentioned in Chapter 1, that fluctuations can also prevent equilibrium states from being realized, physically, so I disagree with this notion.

My own view is that, in most cases, equilibria are special kinds of reversible processes[2] and that static processes are not all reversible, which is

---

[2] A curious situation occurs in considerations of equilibrium theory used to describe observations of piezo-electric and piezo-magnetic effects, as is noted by Ericksen [59]. There is agreement that these are nondissipative processes. However, there are in the literature two contradictory rules for transforming electromagnetic variables. Depending on which one accepts, one of these kinds of effects is not reversible, but the other one is. So, either way, we have an exception to the usual idea that nondissipative processes are the same as reversible processes,

subtle. From Chapter 1, reversible processes are of some importance in the determination of entropy changes, so it is important to know about subtleties pertaining to them. Even if one thinks that they represent limits of processes which cannot be realized, it cannot hurt to know something about what is being approximated. Also, there are theories which treat all possible processes as reversible. Ideas involved are also useful for understanding some other parts of thermodynamics so this deserves some discussion. However, before turning to this, I will make some comments about equilibria which are a little unconventional but are, I think, sound.

In the nineteenth century, it was the view that, in a mechanically isolated system, which may be thermally isolated or in contact with a heat bath at a fixed temperature, equilibrium would be eventually attained and maintained. Expressed one way, some equilibrium configuration would serve as an attractor, other processes approaching it as time increased. Equilibrium theory evolved as a technique for locating such attractors and has proved to be rather successful for this purpose. With the development of the theory of Brownian motion, now called fluctuation theory, it became clear to at least some that this old picture is in error. The fluctuations occur in a somewhat random fashion, keeping the system from attaining equilibrium. Statistically, there is no real limit to the size of the fluctuations possible although, loosely speaking, the probability of occurrence decreases rapidly with size. Still, the attractors determined by the old methods maintain their status as attractors, at least for what seems to us to be long periods in terms of human experience, and it is for this reason that we still take them seriously. Now, experience also indicates that some nonequilibrium processes, for example steady state oscillations, can play a similar role as attractors. However, attempts to devise general thermodynamic techniques for locating such attractors have not been very successful. For special kinds of theories and problems, it may be provable that some measure of dissipation is minimized or maximized by attractors. Understandably, workers have tried to extrapolate such criteria to other kinds of systems with results which may be good or bad, depending on the system. So, I do not feel comfortable in recommending any such criterion. As was mentioned in Chapter 1, there are various other approaches to stability theory, techniques for locating attractors which are not based on thermodynamics. As I see it, if the thermodynamic theory of equilibrium seems to be noticeably different from nonequilibrium theory, it is at least partly because it includes a rather successful stability theory. Otherwise, equilibrium theory deals with a relatively simple subset of processes, and, certainly, it is easier to find constitutive equations, etc., which are capable of dealing with this

---

as I interpret this. Obviously, it would be desirable for physicists to make better sense of this, but as far as I know, no one has.

subset. I think that it is clear from our studies that it is not always so easy to deal with equilibria.

Now, returning to reversible processes, the first step is to consider the variables involved in describing the system of interest and to decide how each of these should be transformed under time reversals. Although not to be discussed here, the Onsager–Casimer relations do have a prominent place in some of the literature on irreversible processes. As is emphasized by Meixner [60], the time reversal transformations play an important role in these. As we commonly understand time, there is no way to cause it to run backwards. However, workers use various ideas to decide what is best. Generally, it is a matter of deciding whether one should or should not reverse the sign of the variable in question. For, say, mass, this is something we always regard as positive, so it is natural to assume that its sign should not be reversed. It is a common notion that what we regard as the more basic equations should transform to equations of the same form. For our theory of bars, this would include the equation of motion (2.3.8), in particular. Let overbars denote the transforms, and assume, as is customary, that position coordinates are unchanged under transformation. Then, for the bars, we obviously have

$$\bar{x} = x, \qquad \bar{y} = y, \qquad \bar{\dot{y}} = -\dot{y}, \qquad \bar{\ddot{y}} = \ddot{y}, \text{ etc.} \qquad (10.3.1)$$

In order that (2.3.8) should transform as an equation of the same form, it is then clear that we should take

$$\bar{f} = f, \qquad \bar{\sigma} = \sigma. \qquad (10.3.2)$$

Where forces are encountered in other contexts, they are to be transformed in a similar way. Generally, the equation representing the first law should retain its form. Power should transform as force multiplied by velocity, i.e. have its sign reversed, from which it is clear that we should take

$$\bar{E} = E, \qquad \bar{P} = -P, \qquad \bar{Q} = -Q. \qquad (10.3.3)$$

For our bar theory, this implies that

$$\bar{r} = -r, \qquad \bar{q} = -q, \qquad \bar{\varepsilon} = \varepsilon \qquad (10.3.4)$$

and, by general consent,

$$\bar{\theta} = \theta, \qquad \bar{\eta} = \eta \Rightarrow \bar{\phi} = \overline{\varepsilon - \theta\eta} = \phi. \qquad (10.3.5)$$

For our theory of bars, this is enough to cover all variables of interest. For another theory, involving different variables, you may need to consult a more experienced worker to determine what are the accepted rules.

The next step is to define the reverse of a process, which may or may not be in the set of possible processes. A process may be reversible for some periods of time and not for others, so it is convenient to define the reverse for any interval of time for which the process is defined. Schematically, let $\alpha(t)$ denote a process which is defined for $t_0 \le t \le t_1$ at least, along with all the responses such as energy, stress, heat flux, etc., which are involved in the theory under consideration. I will call this collection a superprocess. Define the reversed time variable by

$$\bar{t} = t_0 + t_1 - t \Rightarrow t_0 \le \bar{t} \le t_1. \tag{10.3.6}$$

Making this change of variable, we then obtain the quantity

$$\tilde{\alpha}(\bar{t}) = \alpha(t_0 + t_1 - \bar{t}). \tag{10.3.7}$$

Now, apply the time reversal transformations to the variables represented by α, to give

$$\alpha_R(\bar{t}) = \overline{\alpha}(t_0 + t_1 - \bar{t}). \tag{10.3.8}$$

Now, consider $\bar{t}$ to be the usual time, running forward from t_0 to t_1. With this interpretation, it makes sense to ask whether $\alpha_R(\bar{t})$ is in the set of possible processes. When it is, we say that the process $\pi(t)$ involved is a *reversible process* during the time interval considered. Summarizing this, we have

$$\text{process } \pi(t) \text{ reversible} \Leftrightarrow \alpha_R(\bar{t})$$
$$\text{describes a possible superprocess,} \tag{10.3.9}$$

it being understood that this refers to a time interval of the kind described above. As is perhaps obvious, a process which is not reversible for some time interval is called irreversible for this interval. If no time interval is mentioned, a "reversible process" is a process which is reversible for all times for which it is defined.

For, say, the theory of rigid bars discussed in Chapter 3, a number of variables were introduced. For present purposes, it is better to include too many in α than not enough, to cover all relevant transformations. A reasonable choice is

$$\alpha = (\theta, \phi, \eta, q, r). \tag{10.3.10}$$

It would do no harm to include ε, but its transformations, etc., are implied by variables already included. Similarly, one could add x. It is understood that constitutive equations of the kind discussed in Chapter 2 apply so that, given a function $\theta(x, t)$, suitably defined for $0 \le x \le L$ and a time interval $t_0 \le t \le t_1$, we can calculate ϕ, η, and q as functions of x and t. As was mentioned in Section 10.2, there can be differences of opinion as to what is to be assumed about the 'source term' r, in deciding the set of possible processes. To be definite, we will allow it to be assigned arbitrarily, and consider the processes to be smooth, taking the same view of processes that we did in Chapter 2. Thus, the problem of satisfying (2.1.8) is made trivial; we take $r(x, t)$ to be the function given by

$$r = \dot{\varepsilon} - \frac{\partial q}{\partial x} = \theta \dot{\eta} - \frac{\partial q}{\partial x}. \tag{10.3.11}$$

Altogether, this gives the procedure for determining the processes $\pi(t)$. Now, using (10.3.8) and the transformation rules, we calculate the reverse of a process $\pi(t)$, which gives

$$\theta_R(x,\bar{t}) = \theta(x, t_0 + t_1 - \bar{t}),$$
$$\phi_R(x,\bar{t}) = \phi(\theta_R),$$

$$\eta_R(x,\bar{t}) = -\frac{d\phi}{d\theta}(\theta_R),$$

(10.3.12)

$$q_R(x,\bar{t}) = -k(\theta_R)\frac{\partial\theta_R}{\partial x},$$

$$r_R(x,\bar{t}) = \theta_R\frac{\partial\eta_R}{\partial t} - \frac{\partial q_R}{\partial x}.$$

For simplicity, we have used Fourier's law for q, although it is not hard to deal with the general case. Now, for the reverse to be a process, q_R should also be what we calculate using the accepted constitutive equation for heat flux

$$q_R = k(\theta_R)\frac{\partial\theta_R}{\partial x},$$

which disagrees with (10.3.12) unless either $k = 0$ or θ_R is independent of x or, equivalently,

$$k = 0 \quad \text{or} \quad \theta = \theta(t).$$

(10.3.13)

If (10.3.13) holds, then r_R, given by (10.3.12) also satisfies (10.3.11) and such processes are reversible processes. You may be convinced that you can safely neglect the effects of heat conduction in particular cases of interest, accepting $k = 0$ and the ideas concerning r, etc. Then, for you, every possible process is reversible. On the whole, it seems more realistic to regard such processes as irreversible if θ depends on x. Static processes then become irreversible unless θ is constant.

Of course, we can choose interesting subsets of processes by making realistic assumptions about r, (10.3.11) thus becoming a nontrivial equation. Workers who prefer this view will then infer that rather different conditions are needed for processes to be considered reversible. For example, if we insulate the sides to make $r = 0$ and assume $k > 0$, then it is easy to check that these processes are reversible only when θ is constant. If, instead, you assume r to be given by Newton's law of cooling (2.1.17), then, of course, $\theta(x, t)$ must satisfy (10.3.11) for this choice of r and whatever you wish to assume for the heat bath temperature $\theta_B(t)$.

In this case, a calculation gives

$$r_R(x,\bar{t}) = -\alpha[\theta_B(t_0 + t_1 - \bar{t}) - \theta_R(x,\bar{t})].$$

(10.3.14)

From the most obvious way of looking at it, the sign reversal implies that r_R is not given by Newton's law of cooling unless $r_R = 0$, implying that

$\theta(x, t) = \theta_B(t) \Rightarrow q = 0$. Then, (10.3.11) will not be satisfied unless η is independent of t, which case will be true only if $\theta = \theta_B = $ const. So, except in these trivial cases, the processes considered are judged to be irreversible. This is a common view. However, one could modify this argument to allow some such processes to be considered reversible. Suppose that θ depends only on time. Take the view that, in defining the set of possible processes, $\theta_B(t)$ can be considered as any reasonable smooth, positive function of time. Let $\bar{\theta}_B(t)$ denote that associated with the particular process considered, with bar temperature $\theta(t)$ independent of x. Suppose that, for the times considered,

$$2\theta(t) > \bar{\theta}_B(t) > 0. \tag{10.3.15}$$

Then, another possible choice of a function $\theta_B(t)$ is given by

$$\hat{\theta}_B(\bar{t}) = 2\theta_R(\bar{t}) - \bar{\theta}_B(t_0 + t_1 - \bar{t}) \tag{10.3.16}$$

and, with it, we have

$$r_R(\bar{t}) = \alpha[\hat{\theta}_B(\bar{t}) - \theta_R \bar{t}]. \tag{10.3.17}$$

This fits Newton's law, enabling us to regard the reversal as a possible process which is then reversible. In Section 2.2 we considered an example of such a process, there taking a rather conventional view of how it relates to some common thermodynamic experiments.

While these ideas of reversibility are not equivalent, they do have something in common which seems to me to be of some importance, conceptually. It is at least a common intuitive prejudice that irreversibility is associated with dissipation. When the Clausius–Duhem inequality applies, it is commonly accepted that the combination of terms involved provides a measure of dissipation: call it Δ. For our bar theory, this is given by

$$\Delta = \frac{d}{dt} \int_0^L \eta \, dx - \int_0^L \frac{r}{\theta} \, dx - \frac{q}{\theta} \Big|_0^L \tag{10.3.18}$$

and, as will be discussed later, a rather simple expression is used for three-dimensional theories. The usual assumption is that $\Delta \geq 0$ for all processes. For any one-dimensional theory to which this assumption applies, one can use the basic definition of reversible processes to show that

$$\Delta = 0 \quad \text{for all reversible processes.} \tag{10.3.19}$$

Then, the inequality reduces to an equation enabling us to relate changes in entropy to the other terms involved. This provides a basis for experimental determinations of entropy changes whenever we can obtain data enabling us to estimate the other terms. Now reconsider the situation discussed in Section 2.2. If we are willing to assume that Newton's law applies and accept the notion of reversibility indicated by (10.3.15), (10.3.16) and

(10.3.17), we have a reasonable basis for determining changes in entropy from measurements of the bar temperature $\theta(t)$ and heat bath temperature $\theta_B(t)$. For this, it is not necessary that the difference between these temperatures be very small, or that they vary extremely slowly with time. From this view, in Section 2.2, we relied on an old idea mentioned in Section 1.2, that the Clausius–Planck inequality should reduce to equality for reversible processes. Actually, when it applies it defines a different measure of dissipation,

$$\Delta' = \frac{dS}{dt} - \frac{Q}{\theta_B}, \qquad (10.3.20)$$

with $\Delta' \geq \Delta$, typically. What we are discussing are cases where $\Delta' > \Delta = 0$. Commonly, thermodyamicists interested in irreversible processes use Δ as a good measure of dissipation. From this, I infer that what is faulty is the notion that the Clausius–Planck inequality should reduce to equality for reversible processes. It did evolve at a time when it was habitual to assume that the temperature in a system matched that in the heat bath and, of course, this is not true in the example considered.

When the Clausius–Duhem inequality applies, any processes for which $\Delta = 0$ may reasonably be considered in designing experiments to measure entropy changes. There seems to be some prejudice that only reversible processes have this property. I do not see how to make a compelling and general argument for this and the footnote on page 149 seems to imply that there are exceptions.

There is another matter. When we first studied some equilibria, we considered the possible static processes in calculating relevant thermodynamic potentials. Later, we took shortcuts, assuming that the system temperature was a constant matching that of the heat bath. What we were then doing was restricting our attention to reversible processes of a static kind. Had we done this from the start, we would have missed something, namely inequalities involving specific heats. However, the simplified theory is in more common usage and performs fairly well. With it, the theory neatly fits in as a chapter in the thermodynamics of reversible processes, which also includes old theories of heat engines. Conceptually, this is a clearer way of understanding the strategy, I think. That is, find a way to calculate the relevant thermodynamic potential for reversible processes and determine which of these qualify as equilibria. From this view, we really should have added in any possible time-dependent reversible processes in the problems considered, and I have noted examples of problems where it would be necessary to do so. I will leave it to the reader to consider the possibility that this may have some effect on any conclusions we have drawn and what difference it would make if the word "reversible" were replaced by "nondissipative." Of course, there is no obvious reason to exclude irreversible processes, for which we can calculate the relevant potential, and our experience indicates that we may then deduce some additional conditions. From this view, what seems to me to be poorly motivated is the practice of allowing static reversible processes and nothing else. In this, we are following in the footsteps of Gibbs. I have great admiration for him and his work but do not think it wise to follow anyone blindly.

10.4   Cyclic Processes and Cycles

Early ideas about thermodynamics developed from thinking about ancient engines which operate in a cyclic fashion, so it is understandable that cyclic processes do play an important role in the subject. What would seem to me to be the most obvious definition of a cyclic process is one for which the associated superprocess $\alpha(t)$ is defined for all time and is periodic with some period $T > 0$ so that

$$\alpha(t + T) = \alpha(t). \tag{10.4.1}$$

Assuming this, one could define a cycle as the part of a process occurring during any time of duration equal to the period, say

$$\alpha(t), \quad t_0 \leq t \leq t_0 + T \Rightarrow \alpha(t_0 + T) = \alpha(t_0). \tag{10.4.2}$$

One can object to this on the grounds that no real process continues for all time. At least according to some cosmological theories, even our universe has not been in existence for all time. Otherwise, I see no objection to regarding such as examples of cycles. However, in practice, this does not often well describe what is meant by a cycle. Often, as above, a cycle $\mathbf{C}$ is considered to be only part of a process $\pi(t)$, defined over some time interval $[t_0, t_0 + T]$. Roughly, the idea is that

$$\pi(t), \quad t_0 \leq t \leq t_0 + T \text{ is a cycle} \Leftrightarrow \pi(t_0) \doteq \pi(t_0 + T), \tag{10.4.3}$$

where $\doteq$ is to be read "is physically equivalent to." One might stretch the notion of cyclic processes in a similar way, but it is the cycles which are really used in most of the theory. For example, consider one of our thermoelastic bars at rest under no loads, in equilibrium. This could be viewed as a static process, satisfying (10.4.1) for any choice of T. With usual ideas of Galilean invariance, it could instead be travelling with a constant velocity, with the same stretch and temperature. Then, its position $y(x, t)$ would not satisfy (10.4.1). However, if we restrict the process to some interval, we may well regard this as a cycle. In this, there is the implication that, at the beginning and end of a cycle, the related superprocess should have values consistent with the notion of equivalence.

   In discussions relating to existence of energy and entropy, arguments employing cycles are used. Logically, one cannot include things not known to exist in the associated superprocess. So, one considers the related superprocess as consisting of the remaining quantities and makes the judgment about equivalence based on ideas about these only. With this understood, we describe the equivalence by saying that, at the end of a cycle, the system has returned to its original state. When entropy and energy are well-defined, it is understood that their values should be the same at the beginning and end of a cycle.

Necessarily, I think, the notion of a cycle has no precise mathematical meaning. For this, it would be neater to use something similar to (10.4.2). However, the experience is that it is better to have a more flexible concept to deal with a variety of quite different physical situations. In part, thermodynamicists deal with this by trying to formalize the intuitive notion of "state," as is to be discussed shortly.

There is some interest in the design of solid state engines which employ shape memory alloys, involving cycles during which the solid occurs at some times as Austenite, at other times as Martensite. Since engines should do some useful work, the interest is in cycles such that

$$\delta W = \int_{t_0}^{t_0+T} P \, dt < 0, \tag{10.4.4}$$

wherein t_0 and T have the meaning indicated above. The efficiency of such engines, sometimes called the motive efficiency of such cycles, is of interest. To define this, one introduces δQ^+, the total heat absorbed by the system in a cycle. For this, one integrates Q over the set of times for which Q is positive to get a quantity δQ^+ or, equivalently,

$$\delta Q^+ = \frac{1}{2} \int_{t_0}^{t_0+T} (|Q| + Q) \, dt. \tag{10.4.5}$$

Then, the traditional formula for the efficiency e is

$$e = -\frac{\delta W}{\delta Q^+}. \tag{10.4.6}$$

Some engines built employ wire loops which, during a cycle, undergo changes in bending, probably along with changes in stretch. Any mathematical model capable of describing such phenomena and the relevant phase transformations is certainly more complicated than any we have studied. Some possible engines are sufficiently simple to be modelled as thermoelastic bars. Such an analysis is offered by Wollants et al. [uu]. Of those I have inspected, this is one of the better treatments of this kind.[3] If one considers cycles wherein heat is absorbed at just one temperature and is

---

[3]Bad notation and terminology used by these authors may be confusing. For example, the last term in their equation [1] does not make sense because ΔV is defined only on a curve: put the Δ outside the parentheses. Bad terminology is used in referring to adiabatic cycles, for example, which does not mean cycles for which $\delta Q^+ = \delta Q = 0$.

emitted at one other, it is generally believed that Carnot cycles are most efficient.[4] They concur with this but consider other possibilities.

Concerning the basic definition of efficiency, engineers do sometimes modify it. For example, if one is making some good use of what might otherwise be waste heat, one is, by a reasonable interpretation of the word, increasing efficiency. So, it is reasonable to modify the definition to take some account of this. It is worth bearing in mind that, in the latent heats associated with phase transitions, we have what might be regarded as heat sources of this kind and the above authors do consider this.

10.5   States

As was mentioned before, there is general agreement that some notion of states is important in considerations of cycles, energy and entropy, at least. With this is associated the idea that a nonequilibrium process determines states but various different processes can, occasionally, determine the same states. To some degree, concepts of this kind are inherent in classical ideas of causality. That is, to calculate anything physically relevant at one particular time it should not be necessary to know anything about the nature of the process at any later time.

More often than not, when workers speak of states they have in mind what I have called static processes, or something very much like it. It is a reasonable enough idea that one can evaluate a process at a particular instant, to obtain some function which would fit such a description. So, for our thermoelasticity theory of bars, we get the pair $[y(x), \theta(x)]$ from a process $[y(x, t), \theta(x, t)]$, by taking instantaneous values of the latter. Since adding a constant to y is commonly regarded as a trivial change, most would instead use the pair $(\partial y/\partial x, \theta)$ as a description of state. From what we ended up with as constitutive equations, it is easy to check that, given a state, we can calculate the entropy and associate it with a process, as we did previously. According to the first law, we should also be able to calculate the energy but this is more difficult. That is, given our state, we can calculate the total internal energy but not the kinetic energy. As I see it, the best way to fix this is to admit that the notion of state should be generalized to include instantaneous values of the velocity, but I concede that this is not such a popular idea. One trick I have encountered is to consider the "energy" in the first law to mean "internal energy," and to consider power

---

[4]Carnot cycles are discussed in some detail and carefully by Truesdell and Bharatha [1]. From their treatment, it is clear that the 'general belief' tacitly accepts some assumptions. Problems involving phase transitions involve various quirks, so it is wise to think carefully about the possibility and nature of associated Carnot cycles before hastily accepting the belief. Also, these writers include a collection of inequalities useful for analyses of efficiency.

as including the negative time derivative of kinetic energy. This will give an equivalent equation for the theory considered, but, physically, one then needs to revise statements accepted about mechanically isolated systems, for example. As one investigates more sophisticated theories, those of plates for example, one also encounters rotational kinetic energies. If we adopt the indicated viewpoint, these too should be shifted over to give contributions to power. It is reasonable to think that some part of the energy could always be calculated by a formula used for statics. Shift any remainder into power and you have taken care of the first law formally. I do not think that many would be content with this way of interpreting the first law. Other workers do acknowledge that the energy is to include kinetic energy, put in explicitly, but insist that internal energy is to be determined by those static states. As far as our thermoelastic theory is concerned, this is the case.

By playing with the theory of thermoelastic bars, regarding f and r as functions which can be assigned arbitrarily, one can construct processes which are, I think, worth bearing in mind. Consider an unloaded bar, moving with constant velocity in an equilibrium configuration for some constant values of λ and θ, with $t \leq 0$. For $t \geq t_1 > 0$, it again has these values of λ and θ but it is now at rest. I will not expend the ink to prove it, but it is not hard to show that one can find various processes meeting these conditions, adjusting from one to the other fairly smoothly in the interval $[0, t_1]$. As determined by the pair $(\partial y / \partial x, \theta)$, the equilibrium state does return to its original value at time t_1. As I interpret what some writers say that they mean by a cycle, this is one. However, although the internal energy returns to its original value, the energy does not because of the difference in kinetic energy. By our interpretation this process violates the first law which will be discussed later. According to this law the process should then be considered to be impossible. If I thought this were what it really implies I would reject this law. There are various ways of fixing the flaw. One is to interpret energy as meaning internal energy. I have made clear that I do not like this approach. Another is to insist that the "state" is to include velocity which is my wont. A third is to acknowledge that kinetic energy is to be accounted for in an explicit way and as this is the only place where velocity occurs, one need not account for it explicitly in a description of "state." If there is some real difference between this and acknowledging that velocity is to be included in the description of state, it is not apparent to me.

It is true that in the old theories of heat engines such questions did not arise because in them no attempt was made to account for kinetic energy. This is the case for various other theories including, of course, Gibbs' theory of static equilibrium which we have used. A pattern of thinking about states developed from consideration of such examples and habits of thought tend to persist.

Often, one finds writers discussing states with the presumption that they can always be described by a finite list of parameters. There is no reasonable possibility for describing a function, such as $\theta(x)$, by a finite number of parameters. However, for theories which are very local, in a sense more or less like that described in Section 2.2, one can introduce local versions of processes or states. For states, it is a matter of considering what information should be needed to calculate energy and entropy densities at a given position and time. For example, for our bar theory, we need

$$\frac{\partial y}{\partial x}, \quad \dot{y}, \quad \theta \tag{10.5.1}$$

and we argued that ϕ and η cannot depend on $\dot{y}$. What Gibbs [7] discussed were various kinds of local theories, so you can look to see what he assumed about constitutive equations for internal energy and entropy densities and we have covered a few examples. If he covered the case of interest to you, copy it. If not, estimate what it is that he would be likely to have done had he considered it. I think it a fair statement that, from this view, one may have in the local state some functions, and sometimes values of their gradients but not higher derivatives, and temperature or entropy gradients are not included. Assume that these same constitutive equations apply to nonequilibrium processes and you have my interpretation of what is meant by the "hypothesis of local equilibrium" or the 'principle of local states' in the literature on irreversible thermodynamics. Various theories of fluids fit this mould, as does thermoelasticity theory. By this literature, I refer to ideas such as are covered in texts by Kestin [2] or De Groot and Mazur [62], for example. In this respect, theories such as are indicated by (10.1.1) are not local enough, it seems.

From remarks made early in the paper by Gibbs [7], I think it clear that he deliberately excluded the consideration of sliding friction and plasticity. So do the aforementioned texts. Unfortunately, phenomena of this kind are common and important in solids. With the concept of state used by Gibbs and in the aforementioned tests it is at least tacitly understood that if a state can be attained once, it can be attained again and again and that, once it is attained, whatever happens thereafter does not depend on how many times it has been attained in the past. Also understood is that one can find many cyclic processes containing any given pair of states, which can also be joined by processes which are reversible or at least processes well approximating this. I appreciate that these are loose statements, but, at least, hopefully, it is enough to provide some intuitive understanding.

Take a paper clip and bend part of it sufficiently to induce what seems to be permanent deformation. Now try to unbend it, to return it to the original "state." You are likely to find that this is at least difficult but you may think that you have succeeded. If so, try to repeat the process. If the wire breaks, or you see some other discernible difference, you have not attained the goal. It is hard to exclude the possibility that someone may succeed by making some clever use of heat treatments, for example. We might agree that we could succeed by melting and re-forming the wire. This is acceptable, if you are willing to construct a theory general enough to include such processes. What seems clear is that, if we want "states" to have such conventional properties, it can be difficult to find many cycles beginning and ending at a particular state. Whether there are any which also pass through another given state is not always so obvious. If anything, finding reversible processes connecting a given pair of states, or good approximations of these, is harder. I will not try to discuss in detail the numerous theories of plasticity which try to deal with thermal effects. For bars, typical theories would use the equation of motion (2.3.8) and energy equation (2.3.12) which we used for thermoelasticity theory, together with a constitutive equation for ε. As before, the idea is to use the energy equation as an equation for θ.

Because of difficulties such as are mentioned above, it is not clear how to make good sense of entropy or states and these concepts seem not to have been very useful. Replacing one constitutive equation for stress is a more complicated strategy for relating changes in stress, deformation, and temperature. It involves two sets of equations with a criterion to determine which applies, which depends on the process. Given the situation, it is not always clear what is to be meant by a cycle, but I do not think that anyone working in this field would object to the notion that there are numerous cycles. This is enough to make it possible to make use of laws of thermodynamics to be discussed later. Perhaps there is some better way to make use of thermodynamics in such situations and different workers do explore different approaches. Not being expert in this, I do not feel comfortable in trying to say more. In any event, plasticity is a very common phenomenon in metals, in particular, and it is at least awkward to fit it into a general scheme of thermodynamics. Rather general theories of thermoplasticity are proposed by Green and Naghdi [63]. Somewhat similar difficulties are encountered in theories of damage such as are covered in the review by Chaboche [64], for example.

Rather different difficulties are encountered in other common kinds of solid materials, for example, in the elastomers, the various kinds of rubber. Here, equilibrium theory such as we used for balloons and sheets can be combined and considered as different ways of using three-dimensional thermoelasticity theory. This does have a rather wide range of applicability and it has been used quite successfully to master numerous phenomena. However, for more time-dependent processes, viscoelastic effects are important and these do not fit comfortably into the pattern treated in texts on irreversible thermodynamics. Actually, if you look into the literature on observations of those equilibria, you will find some workers acknowledging that what has been seen is not really static but is changing very slowly with time. How slowly depends on the nature of the experiment and the type of rubber. So, there are possibilities for minimizing such effects, perhaps getting them sufficiently small to be hidden by experimental errors for the time scales involved in the observations. Still, it is hard to avoid the impression that equilibrium theory is being misused but with some success.

Some workers like to think in terms of a "rubber plateau," involving time scales neither too short nor too long, over which relaxation processes slow down enough to obtain something that looks much like equilibrium, in an experiment designed to produce a static response. For example, one view is that if one clamped the edges of a sheet to produce unequal stretches, one could produce results similar to those discussed in Chapter 6, involving some shear stresses, for a time which may seem very long. However, the notion is that those shear stresses will eventually relax to zero. To see this, one might need to observe them for times better measured in months or years. Think of silly putty and you see behavior something like this but speeded up considerably. I do not regard this as an unreasonable point of view. It does suggest that true equilibrium theory might be that appropriate for a fluid and we are using something quite different from this for the nonequilibrium processes occurring on the plateau. Intuitively, we are then not close to real equilibrium. Put one way, we seem able to use Gibbs' ideas to find some processes which, for a limited time only, serve as attractors. Given this, there is some reason to believe that it should be feasible to use more accurate viscoelasticity theory to account for the slow changes and to associate with this a means of calculating energy and entropy. Reasonably, this should not differ much from what we use for rubber elasticity on the somewhat vaguely defined rubber plateau. There are others who prefer to believe that, instead of reaching such a plateau, we are getting close to real equilibrium, better described by theory more like the rubber elasticity we have used. Also, there is a range of opinions about whether it is feasible to make sense of entropy as it pertains to viscoelasticity theory and, if so, how it should be done. The situation is not so clear that reasonable persons cannot disagree about such matters, as I see it. What is involved is a kind of theory too complicated to cover in an elementary way. However, I will try to give some indication as to why such theories involve some unusual difficulties.

More often than not, nonlinear theory is needed to describe phenomena of interest. However, linear one-dimensional theory is much simpler and is useful for describing some phenomena, for example, small vibrations. Even here, the basic difficulty appears. Often, the interest is in behavior in shear, so we will consider the interpretation in terms of the shearing of plates. The difficulties are more associated with the mechanics and there is some question concerning the thermodynamics, so I will consider purely mechanical theory. When I first came into contact with such work, it was common for workers to draw pictures of springs and dashpots linked together in some fashion, using this to motivate considering equations of the form

$$\sum_{m=0}^{M} a_m \frac{\partial^m \tau}{\partial t^m} = \sum_{n=0}^{N} b_n \frac{\partial^n \gamma}{\partial t^n}, \qquad (10.5.2)$$

as a substitute for a constitutive equation for τ. Here, as before, τ and γ are, respectively, the shear stress and strain. This is to be used in conjunction with the equation of motion (2.4.5). The a's and b's are constant, related to spring constants and dashpot viscosities in the picture, from which you can infer their algebraic signs. Typically, M and N differ by one at most. If one tries to match data on small vibrations, the experience is that, if you try to get by with small values of M and N, you can only cover a very small range of frequencies with resonable accuracy. As you try to increase the range of frequencies covered, you need to increase M and N and there seems to be no upper limit to this. In this respect, such models are not very satisfactory. However, workers do still use them or nonlinear generalizations of them, to analyze some kinds of phenomena. Setting all derivatives equal to zero, one gets τ proportional to γ. This gives for the static elastic shear modulus, the value b_0/a_0. If you conceive the material to be a fluid, set $b_0 = 0$. Otherwise, any difference between fluids and solids is more quantitative, more associated with values of the constants estimated from experiment. In another respect, such models are at least awkward. Typical relaxation experiments involve suddenly imposing a deformation or force and holding it. So the time derivatives involved are, initially, very large, perhaps infinite.

With such experiences, workers were encouraged to consider a different model proposed in 1874 by Boltzmann [65]. It replaces (10.5.2) by an equation of the form

$$\tau(x, t) = \mu_\infty \gamma(x, t) + \int_{-\infty}^{t} h(t - \sigma)[\gamma(x, t) - \gamma(x, \sigma)] \, d\sigma, \qquad (10.5.3)$$

where μ_∞ is the equilibrium shear modulus and $h(t - \sigma)$ is a function depending on the material. It is considered to decrease quite rapidly as $\sigma \to -\infty$. In the sense indicated, the material remembers all deformations occurring in the past but not as well those that occurred a long time ago. Philosophically, it is not so pleasant to think that one must know these ancient deformations but many workers find ways to live with the idea. Also, there are the obvious practical difficulties in knowing what to assume about this. If you prefer, you can think of starting with having the plate undeformed ($\gamma = 0$) up to a certain time, which alleviates the problem but does not completely eliminate it. Some relevant experimentation is covered by Ferry [66]. Three-dimensional linear theory is discussed by Leitman and Fisher [67], for example.

A stress relaxation experiment involves starting with a sample which seems to be in equilibrium, with $\gamma = 0$ for $t < 0$, suddenly imposing a constant shear strain γ_0 and holding it, then measuring τ as a function of time. The aim is to obtain an experimental determination of

$$\mu(t) = \tau(t)/\gamma_0. \qquad (10.5.4)$$

FIGURE 10.1. Hypothetical form of response in stress-relaxation, for an elastomer, with the idea that, eventually, the shear stress relaxes to zero.

Usually, no attempt is made to measure the temperature of the sample or the heat transferred. For such reasons, these have no clear status as thermodynamic experiments. Because of inertial effects, the initial jolt produces waves more or less like the shock or stress waves treated in Section 7.1, so one cannot really attain the constant value of γ_0 immediately. However, these effects seem to disappear rather quickly so one can begin to get meaningful measurements of $\mu(t)$ after a short lapse of time. At long times, one is limited more by the patience of the observer. By a simple calculation, ignoring inertial effects, we should have

$$\mu(t) = \mu_\infty + \int_{-\infty}^{0} h(t - \sigma) \, d\sigma. \tag{10.5.5}$$

With the change of variables $z = t - \sigma$, this becomes

$$\mu(t) = \mu_\infty + \int_{t}^{\infty} h(z) \, dz. \tag{10.5.6}$$

Granted that $h \to 0$ fast enough as $z \to \infty$, we will have

$$\mu_\infty = \lim_{t \to \infty} \mu(t), \tag{10.5.7}$$

and

$$\mu'(t) = -h(t), \tag{10.5.8}$$

which gives a simple physical interpretation of the function h and a way of determining it for a range of values of the argument. A worker who likes to think of elastomers as fluids would want to have $\mu_\infty = 0$ and may picture $\mu(t)$ as indicated in Fig. 10.1 for such "solids."

It should have a long, almost flat part representing the rubber plateau, with $\mu \to 0$ as $t \to \infty$. A worker who watches only long enough to encounter the plateau might reasonably extrapolate to obtain some nonzero value of μ_0 and dislike the notion that it is to be viewed as a fluid. It may be valuable for someone to collect all the evidence bearing on this and perform a careful critique of it but, as far as I know, this has not happened.

So, this gives a crude picture of the theories of concern. The first version is in a sense local, but hardly local enough to fit the pattern of very local theories commonly treated in texts on irreversible thermodynamics. The pictures of springs and dashpots cannot be taken too seriously. However, in it, there is the suggestion that, at least temporarily, it is possible to store some energy in the springs, even if we are dealing with a fluid and this is at least intuitively consistent with the "elasticity" which they exhibit. Certainly the second version is far from local, in the temporal sense.

Now, if we ignore the inertial effects, we can calculate the initial elastic modulus as

$$\mu(0) = \mu_\infty + \int_0^\infty h(z)\,dz. \tag{10.5.9}$$

In this quick stretch, the body is behaving like a linear elastic body and we can use this idea to make an estimate of the work done per unit volume, δW,

$$\delta W = \mu(0)\gamma_0^2/2. \tag{10.5.10}$$

In cases like this, it is rather commonly accepted that there is too little time for appreciable heat transfer to take place, particularly in the elastomers which are not good conductors of heat. So, by the first law, this also gives an estimate of the energy density, really the internal energy density since we are neglecting inertial effects. With elastomers and various other high polymers, like silly putty, which are more obviously fluids, the initial jolt should be close to a reversible process, intuitively. Thus, it makes some sense to regard this as an isentropic process and assign to this initially stretched configuration the value of the entropy occurring before stretching. As a personal matter, I do not think that these ideas are far wrong, and, as far as I am concerned, they make unreasonable the idea that you can use the hypothesis of local equilibrium. To avoid some arguments, consider something like silly putty which is obviously better regarded as a liquid. Such liquids are reasonably considered as incompressible and besides, the shearing motions do not produce volume changes according to linear theory. For equilibrium, the internal energy should then not depend on anything but entropy. If the entropy is fixed, so is the internal energy. There is then no way to explain the change in internal energy inferred from (10.5.10) for such liquids. If you accept the idea indicated in Fig. 10.1, the same argument applies to elastomers. If you do not, then the problem is to explain why the hypothesis should work for viscoelastic solids but not viscoelastic fluids. I consider this to be hopeless, so I reject the hypothesis. Various workers have come to this conclusion, quite possibly for different reasons. If you agree and also accept the viewpoint of Tisza, among others, mentioned in the Preface, then the problem of defining energy and entropy is easy. It is impossible, so forget it. Of those I know in this business, no one seems completely happy with this resolution. Some may well be prepared to believe that this is a possible conclusion but they would need to see some more compelling argument to accept it.

For the elastomers, as we get well into the plateau or, by the other view, close to equilibrium, the temperature is likely to be again at least close to the ambient. Either way, we are in a regime where, by common consent, thermoelasticity theory applies, at least to a fairly good approximation. Said differently, the notion that energy and entropy have meaning in this regime seems to be generally accepted. At least approximately, we can here use thermoelasticity theory to calculate them. For the situation considered, this leaves us without an obvious way of calculating these quantities in the regimes where $\mu(t)$ is changing more rapidly with time. This is one example of a regime for which experts seem not to have come to a very good agreement concerning how best to deal with these quantities. There are several reasons for this. As was mentioned earlier, some might hold to the view that one can only use those equilibrium states excluding the possibility of defining entropy, etc. Also, as was mentioned, the basic ideas of such theory do involve some philosophical and practical difficulties and workers can have rather different ideas about how best to deal with them. For example, we mentioned the possibility of considering just one standardized initial history for a material, or allowing for more, and I have encountered some differences of opinion about this. Also, nonlinear theory of this kind is very complex, making it rather easy for workers to disagree about how reasonable it is to make rather technical assumptions which seem to be needed to perform mathematical analyses relating to general theories of this kind. For such reasons, it is understandable that an argument which seems convincing to one expert can be considered to be unconvincing to another. As best as I can assess the situation, Coleman and Owen [68] seem to have come closest to giving a satisfactory treatment of energy and entropy for such theories. In this, the distinction between states and processes gets blurred, with states becoming histories of deformation, etc. The idea of cycles is modified by introduction of a concept of approximate cycles. With this, the conclusion is that energy and entropy, like stress, do depend on those histories. As might be expected from the fact that this is a complex theory, their treatment is hardly in the elementary category.

For theory of this kind, one can find some things which do fit the description of cycles. For the purely mechanical theory given by (10.5.3), one can use the idea represented by (10.4.1) and (10.4.2) to generate some cyclic processes. That is, it is not hard to show that

$$u(x, t+T) = u(x, t) \supset t \Rightarrow \tau(x, t+T) = \tau(x, t). \qquad (10.5.11)$$

It is not immediately obvious that this is implied by the other version (10.5.2) but, in practice, workers accept this. Also, the corresponding result holds for at least most nonlinear theories of viscoelasticity of the "memory" kind encountered in practice, including three-dimensional theories. For the mechanical theory, this is enough to justify regarding such processes as cyclic, generating some cycles. Intuitively, I would not be happy with a definition of cycles or cyclic processes which let the stress have different values at the beginning and end of a cycle. Here, it seems difficult to find any very different kinds of cycles which have this property; one could generalize that used slightly, requiring only that $\gamma(x,t)$ be periodic in time. Of course, the mechanical theory should be generalized to include equations governing temperature variations if we are to seriously consider applying thermodynamic ideas to such theories. These periodic processes are not the only processes of interest and one would like to have a much greater variety of cycles to make better use of the laws of thermodynamics. As in the plasticity case, the consensus of opinion seems to be that there is no need to modify conventional equations of motion or energy equations. Simply, it is not a matter of routine to apply thermodynamics when, as here, one is dealing with theories not of the very local kind. To some degree, this explains why Coleman and Owen [67] were motivated to modify the notion of cycles, to obtain more things which could play a similar role. Once one gets away from the equilibrium problems for elastomers, one is into issues too complex to be covered in an elementary treatment of relevant thermodynamics.

These are not the only difficulties encountered in trying to apply thermodynamics to improve our understanding of solid behavior, in particular. What we have available to overcome them are the basic laws of thermodynamics. I thought that it may help to have some inkling of those difficulties encountered in practice before proceeding, since thermodynamicists rarely mention these.

10.6 Laws of Classical Thermodynamics

To some degree, it is a subjective decision as to what should be considered as classical thermodynamics. I and some others prefer to interpret it rather broadly, to at least leave open the possibility of applying the ideas to troublesome cases, such as were mentioned in Section 10.5. Others prefer to limit it to the cases which are much more routine, covered by the hypothesis of local equilibrium, etc.

A rather common classical view as I interpret it is that heat, power and temperature are the most basic entities, in the sense of being most closely related to ordinary experience. More abstract are concepts of energy and entropy. If these are to be meaningful, they, and whatever properties they might have, should be inferred from ideas dealing more with the basic concepts. I believe that it helps to understand what is done if you accept this view, at least for the sake of argument. Basically, the laws declare some things to be physically impossible. At least to some degree, they grew out of experience with ancient engines, so it is not surprising that cycles play an important role in them. Involved are verbal statements which are subject to interpretation, although there is not very much disagreement about this. I will make some comments about this along the way.

The first has some implications for any thermodynamic system capable of undergoing at least one cycle. It reads

For any cycle, beginning at time t_0 and ending at time t_1,

$$\delta W = \int_{t_0}^{t_1} P\,dt = -\delta Q = -\int_{t_0}^{t_1} Q\,dt. \qquad (10.6.1)$$

By general consent, one implication of the first law is that, for a loading device to be considered to be conservative, δW should vanish for every possible cycle. This does exclude some things which may be considered to be dead loading devices. For example, if one maintains a constant torque on a stir rod inserted in a viscous fluid, one could view this as dead loading but, in likely cycles, δW is positive, so this is not a conservative device. Numerous elementary texts discuss how one can use this law to infer the existence of an energy depending on states for sufficiently simple theories of systems. I will not expend the ink to give one more discussion of this kind. However, I will note that, with our theory of thermoelastic bars, this statement would fail if we regard as cycles those in the example mentioned in Section 10.5, where the kinetic energy does not return to its original value. Obviously, the discussion of the first law given in Chapter 1 makes a presumption about existence of energy. Here, this is to be viewed as something to be proven starting from (10.6.1). With the troublesome theories mentioned in Section 10.5, this is one of the nontrivial problems. Various implications of impossibility can be read from (10.6.1). For example, if you wish to obtain some useful work from a system ($\delta W < 0$), you cannot also get the system to emit more heat than it absorbs, (i.e. having $\delta Q < 0$). It is easy to see that, for the efficiency e defined by (10.4.6), the first law implies when $\delta Q^+ \neq 0$,

$$0 \le e \le 1. \qquad (10.6.2)$$

With the second law, the situation is more complicated. Different statements were made by older writers and one finds these restated in different ways in different texts. Some are convinced that these are equivalent and others are not. Serrin's description [69] of the original statements is accurate. His paper begins with a quotation of a statement by Bridgman, a noted physicist:

> *It does seem obvious that not all formulations of the second law can be exactly equivalent.*

I think it pertinent to add that his experimental work[5] turned up numerous interesting facts relating to phase transitions and plasticity phenomena

---

[5]This is covered in Bell's review [13].

in solids. Along with this, he was a serious student of thermodynamics with a strong interest in trying to apply it to plasticity phenomena, in particular. He did appreciate at least some of the difficulties which this involves.

Commonly mentioned in texts is a version of the second law attributed to Clausius. As is discussed by Serrin [68], Clausius proposed two versions. To this, I would add a third. As noted in Section 1.2, Gibbs took a statement by Clausius as a basis, or at least a motivation for his work. Some take it as a motivation for considering (1.2.7) as a statement of the second law, one version not mentioned by Serrin. To take this as basic, it seems to be necessary to believe that for any thermodynamic system covered by classical thermodynamics, entropy can somehow be defined. Presumably, the later of the other two versions mentioned by Serrin is the best formulated. It reads as follows:

(Clausius) *A passage of heat from a colder to a hotter body cannot take place without compensation.* (10.6.3)

Clearly, this recognizes that the intuitive idea that heat flows from hotter to colder bodies is not always true: if it were, it would not be possible to make refrigerators. Some notion of temperature is involved to discern which is hotter. As noted in Section 10.1, I have accepted the assumption that we can always consider temperature to be absolute temperature. Nominally, the first law contains no reference to temperature. However, for the first law to make sense, one needs to be able to assign some numerical value to heat in any particular situation, which means using some physical units of heat. If one looks at the definition of such a unit, say a calorie, one finds that it involves a definite unit of temperature, a degree on the centigrade scale. The other point is, that to compare work with heat, we must use the same units for both. Converting heat data to these is performed by multiplying by a universal constant, which does of course depend on which units are selected. So, in practice, we do accept a universal temperature scale in interpreting the first law.

As at least some interpret (10.6.3), the colder and hotter bodies are to be described as two heat baths at fixed temperatures θ_1 and $\theta_2 > \theta_1$. Some thermodynamic system, allowed to undergo a cycle, is somehow to be in contact with these at some times during the cycle. Any heat transfer is to be between the system and the heat baths. Thus, for example, when the system is in contact with one, any heat absorbed (emitted) by the system is emitted (absorbed) by this heat bath. When the system is in contact with neither, the process must be adiabatic ($Q = 0$), etc. The phrase "without compensation" is taken by some to mean that the net work done on the system during a cycle must vanish ($\delta W = 0$). Then, from (10.6.1), we should also have $\delta Q = 0$. I think it reasonable to relax the restriction as Serrin does [68], for example, using the interpretation that

"without compensation" means $\delta W \le 0 \Rightarrow \delta Q \ge 0$, (10.6.4) where I have used the first law.

To decide for yourself whether such a statement makes sense, a first step is to check whether it does hold for systems familiar to you. If so, you can try to design and build a system which violates it. If you cannot, you acquire some faith that the law is sound, but there is no way to prove this. Let us try one check using the thermoelastic theory of bars discussed in Chapter 2.

For a bar, consider a cycle beginning at $t = 0$, ending at $t = T$, conforming to (10.6.4), so that

$$\delta Q = \int_0^T Q \, dt = - \int_0^T P \, dt = -\Delta W \geq 0. \tag{10.6.5}$$

To bring in the contact with the heat baths, we consider the ends to be insulated, so

$$q(L, t) = q(0, t) = 0 \supset t, \tag{10.6.6}$$

and use Newton's law of cooling, with the assumptions that, for $0 \leq x \leq L$,

$$
\begin{aligned}
r &= \alpha[\theta_1 - \theta(x, t)], & 0 &\leq t < t_1, \\
r &= 0, & t_1 &\leq t < t_2, \\
r &= \alpha[\theta_2 - \theta(x, t)], & t_2 &\leq t < t_3, \\
r &= 0, & t_3 &\leq t < T,
\end{aligned}
\tag{10.6.7}
$$

with α some positive constant, $\theta(x, t)$ being the temperature of the bar. It is not immediately obvious that any such cycles exist, since some equations must be satisfied, and I will not deal with this question. The question is more to check whether any that may exist do conform to this version of the second law. To be definite, by a cycle, we mean a process defined for times in the interval $[0, T]$, with

$$y(x, T) = y(x, 0) \Rightarrow \lambda(x, T) = \lambda(x, 0), \theta(x, T) = \theta(x, 0) \tag{10.6.8}$$

satisfying the equation of motion (2.3.8) with $f = 0$ and the heat equation (2.3.27) with r as prescribed in (10.6.7). One can allow for such things as the shock waves discussed in Section 7.1 but I will leave it to you to determine whether this affects the conclusions we will make. Now, with (10.6.8) we will have, in particular

$$S(T) - S(0) = \int_0^L \eta(\lambda, \theta) \, dx \bigg|_0^T = 0. \tag{10.6.9}$$

Then, from the global form of the Clausius–Duhem inequality (2.3.13), with $x_1 = 0$, $x_2 = L$, we obtain

$$\int_0^T \int_0^L (r/\theta) \, dx \, dt \leq 0, \tag{10.6.10}$$

where we have used (10.6.6). With (10.6.7), this gives

$$\alpha \int_0^L \left\{ \int_0^{t_1} [(\theta_1 - \theta)/\theta] \, dt + \int_{t_2}^{t_3} [(\theta_2 - \theta)/\theta] \, dt \right\} dx \leq 0. \qquad (10.6.11)$$

From (10.6.4), we must also have

$$\alpha \int_0^L \left[\int_0^{t_1} (\theta_1 - \theta) \, dt + \int_{t_2}^{t_3} (\theta_2 - \theta) \, dt \right] dx \geq 0. \qquad (10.6.12)$$

Here the factor α, being positive, can be canceled. Now, with our interpretation of this version of the second law, it should be impossible for heat to be transferred from the colder to the hotter bath under the conditions assumed. That is, briefly,

$$\text{Clausius} \Rightarrow \alpha \int_0^L \int_{t_2}^{t_3} (\theta_2 - \theta) \, dt \, dx \geq 0. \qquad (10.6.13)$$

The question is whether this really does follow from (10.6.10) and (10.6.12), perhaps by using something else deducible from the bar theory considered. For this, it is helpful to note that, for any possible values of θ_1 and θ,

$$(\theta_1 - \theta)(\theta_1^{-1} - \theta^{-1}) \leq 0 \Rightarrow (\theta_1 - \theta)/\theta_1 \leq (\theta_1 - \theta)/\theta. \qquad (10.6.14)$$

Of course, one can here replace θ_1 by θ_2. Thus, using (10.6.11), we have

$$\int_0^L \left\{ \int_0^{t_1} [(\theta_1 - \theta)/\theta_1] \, dt + \int_{t_2}^{t_3} [(\theta_2 - \theta)/\theta_2] \, dt \right\} dx$$

$$\leq \int_0^L \left\{ \int_0^{t_1} [(\theta_1 - \theta)/\theta] \, dt + \int_{t_2}^{t_3} [(\theta_2 - \theta)/\theta] \, dt \right\} dx \leq 0. \qquad (10.6.15)$$

Now, use (10.6.12) to eliminate the first term on the left to give

$$(\theta_2^{-1} - \theta_1^{-1}) \int_0^L \int_{t_2}^{t_3} (\theta_2 - \theta) \, dx \, dt \leq 0, \qquad (10.6.16)$$

and, with $\theta_2 > \theta_1$, this does give the inequality alleged to hold in (10.6.13). So, as interpreted here, we have verified that (10.6.3) does hold in the cases considered. It is not hard to check that we would have encountered a violation had we considered α to be negative. Said differently, we have come close to proving that $\alpha \geq 0$ follows from (10.6.3). To complete the proof, one would need to show that there is at least one cycle of the kind considered. If you are willing to assume this, you have another reason to think that (10.6.3) makes sense as interpreted here. Suffice it to say that this old law is still accepted because it has been tested in numerous ways and found to be trustworthy.

As is also commonly discussed in texts, one can use this version of the second law to deduce a useful inequality, even if one insists on having equality hold in (10.6.4). Consider any system in contact with a heat bath with temperature $\theta_B(t)$. Suppose that the system is capable of undergoing a cycle beginning at time t_0 and ending at time t_1. For the cycle, we then have

$$\int_{t_0}^{t_1} [Q/\theta_B(t)] \, dt \le 0, \tag{10.6.17}$$

this being called the *Clausius inequality*. In various simpler theories, it is automatic that θ_B is also the temperature of the system and some writers do not make a clear distinction between these two temperatures. Also commonly discussed in texts is the fact that, for relatively simple systems, one can use (10.6.17) to deduce the existence of entropy, depending on suitably defined states satisfying the Clausius–Planck inequality (1.2.2). Also for such theories, it is generally easy to show that (1.2.2) implies (10.6.17). In any event, (10.6.17) represents Clausius' contribution to the Clausius–Duhem inequality which we will discuss a little more later. For our bars, another possibility is to have the two heat baths in contact with the bar at the same time. For example, we could insulate the sides so $r = 0$ and put each end in contact with different heat baths, each at fixed temperature. With a suitable control of mechanical conditions, one may attain a static process which is, of course, cyclic. It is hard to see how (10.6.17) could yield any information about this. From such reasoning, it seems clear that the Clausius inequality should not be regarded as the same as the second law, as interpreted above.

Another commonly mentioned version of the second law is Planck's restatement of Kelvin's idea:

(Kelvin–Planck) *It is impossible to construct a machine which operates cyclically and which does nothing but raise a weight and cause a corresponding cooling of a heat reservoir.*

(10.6.18)

Most regard this as equivalent to Clausius' version. Serrin [68] does discuss assumptions which, it seems, need to be accepted, at least tacitly, in establishing equivalence, at least as he interprets the statements. I have nothing worthwhile to add to such discussions, so will leave it at this.

In these versions of the laws, the basis is a relatively simple intuitive idea, structured by conventions which have been established from long experience in using them. In the basic statements, as interpreted here, there is no explicit mention of states. By convention, notions of states have come to be associated with the statements, at least largely because this is a convenient way of describing the energies and entropies which can be deduced using these laws, when we know how to do this. However, there is no implication that such energies and entropies exist for all kinds of thermodynamic systems and I do not believe that they do. The laws have some implications for any system which can undergo at least one cycle. If, by any mode of reasoning, one can decide what are cycles, one can explore this without also having any preconveived notions about states. With various theories of plasticity and viscoelasticity, it is rather clear that one has some variety of cycles and the above versions do have some implications about these. In such cases, the implications may well be less than what would be needed to deduce well-defined entropies at least, but it is not a classical law that this should always be possible. It would not necessarily violate the laws if one assumed that they are well-defined and gave some prescription for calculating them. If you try this, you may well have trouble in convincing others that your ideas are sound.

Also commonly mentioned is a third version, quite different in character, due to Carathéory. First, recall the statements made at the beginning of our discussion of the second law, by Bridgman, and accept that he was familiar with the three. Another good physicist, Pippard [4], takes a rather contradictory view, so I think it worthwhile to explore this. He states what he consideres to be this version,

> (Carathéodory–Pippard) *In the neighbourhood of any equilibrium state of a system, there are states which are inaccessible by an adiathermal process.*   (10.6.19)

It is clear from his discussion that, where I have used the word "adiabatic," he means "adiathermal." He notes that he likes this least, because

> *...it is neither intuitively obvious nor supported by a mass of experimental evidence.*

However, as he puts it,

> *...it leads to the same conclusions as the others....*

Clearly, this view is not easy to reconcile with that of Bridgman. Physicists tend to skip little points of mathematical rigor, so one might reasonably explore what mathematicians have to say about this version, one possibility being the previously mentioned paper by Serrin [68]. Here, we find a rather different statement, viz.

> (Carathéorody) *Consider a system S whose behaviour can be described by a finite number of state variables whose domain is an open set Ω of R^k, $k \geq 2$ (with the usual topology). Let σ_0 be a state of the system (a point in Ω) and let N be any neighbourhood of σ_0 in Ω. Then there must exist in N a point σ with the property that no adiabatic process open to S can start at σ_0 and end at σ, (i.e., σ is inaccessible from σ_0).*

(10.6.20)

Read on, and you will see that Serrin regards this as a very different version, not equivalent to the previous two.

Two things are worth noting. It is more than glossing over points of rigor to fail to mention that those states should be describable by a finite number of parameters, as Pippard does. Also, what is a state σ_0 in (10.6.20) is, in Pippard's version, an "equilibrium state." So, accept that the world of classical thermodynamics is limited to the cases which can be described in terms of states which are, in turn, describable by a finite list of parameters. Further, such states are to fit the description of equilibrium states discussed earlier. From what else he does, this is what I infer to be Pippard's view of what is classical thermodynamics and, in this respect, he has lots of company. Workers who accept this view generally accept the idea that energy and entropy are well-defined as functions of such states. This view is too limited for some, Bridgman for one, and I have declared myself to be with him in this. I have said enough about the problems met in deciding what should be meant by states to indicate that, sometimes, I do not see how to use the third version in cases where the other two can be applied. If I had some conjecture about this, I would return to the previous versions to try to decide whether it made sense. In this respect, those versions are certainly not replaceable by the third version for my purposes. The third version has some merit in providing a basis for simple equilibrium studies such as we have performed, when one does not want or know how to place these in the context of some nonequilibrium theory. However, even here, it is hard to see how one could use this to motivate or justify what we took to be basic definitions of equilibrium. Neither does it seem to warn us against trying to use them for problems involving sliding friction, for example. As I see it, the view adds little to what Gibbs said about equilibrium. For such reasons, I do not find the third version very useful, but you might.

Easier to use, when it applies, is the Clausius–Duhem inequality, which really comes in two versions. We have already discussed this in relation to one-dimensional theories. For three-dimensional continuum theories, the more traditional version goes as follows. Some fixed set of material is considered to be the thermodynamic system. Here, entropy is introduced from the start. Introduce a reference configuration, some possible configuration as a three-dimensional analogue of what we have used for bars and plates. Let Ω denote the region it occupies. With S denoting the entropy, η the entropy per unit reference volume, we have, by assumption

$$S = \int_\Omega \eta \, dv, \qquad (10.6.21)$$

dv being the volume element. Here, to describe Q, we introduce a heat flux vector q and set

$$Q = \int_{\partial\Omega} q_n \, da, \qquad (10.6.22)$$

where $\partial\Omega$ denotes the boundary of Ω, da the element of area and q_n the projection of q on the outward normal to the surface. Then, the Clausius–Duhem inequality is

$$dS/dt \geq \int_{\partial\Omega} (q_n/\theta)\, da, \qquad (10.6.23)$$

where, as usual, θ is the temperature of the material which, in general, varies with position and time. By using the divergence theorem, etc., one can obtain a corresponding inequality which holds pointwise, but I will not belabor this. If you return to Chapter 1 and read the quotation from Gibbs, you will find that rather puzzling integral, in which t denotes

...*the temperature of the part of the system receiving it.* ...

It is not clear what he had in mind but it seems not to be quite the integral occurring in the Clausius inequality (10.6.17); something more like that occurring in (10.6.23) seems to be the case. With the assumption that the quantity on the left vanishes when integrated over a cycle, one gets an inequality somewhat like the Clausius inequality,

$$\int_{t_0}^{t_1} \int_{\partial\Omega} (q_n/\theta)\, da\, dt \leq 0. \qquad (10.6.24)$$

In the above discussion of bars we mentioned the possibility of putting different heat baths in contact with the ends of a bar. Kestin [70] uses a similar idea to conclude that (10.6.23) is deducible from our (1.2.7) when entropy is defined using the hypothesis of local equilibrium. Possibly, one could use a similar argument to deduce that (10.6.24) follows more generally from, say, Clausius' version of the second law. This could be useful in trying to justify the assumed existence of entropy for more general systems by this route, but I am not sure of this.

The other version of this inequality has been frequently used since it was suggested by Truesdell and Toupin (p. 258 of [71]). For bars the quantity r may be thought of as being defined within the bar. The proposal is to modify (10.6.22) in that analogous way, writing

$$Q = \int_{\partial\Omega} q_n\, da + \int_{\Omega} r\, dv, \qquad (10.6.25)$$

replacing (10.6.23) by

$$dS/dt \geq \int_{\partial\Omega} (q_n/\theta)\, da + \int_{\partial\Omega} (r/\theta)\, dv. \qquad (10.6.26)$$

Then, (10.6.24) is modified in the obvious way to

$$\int_{t_0}^{t_1} \left[\int_{\partial\Omega} (q_n/\theta)\, da + \int_{\Omega} (r/\theta)\, dv \right] dt \leq 0. \qquad (10.6.27)$$

It seems less likely that one could produce a satisfactory derivation of this from any of the other three versions of the second law. For such reasons, it is at least awkward to tackle problems related to the existence of entropy with suitable properties, for some systems of this kind, in particular theories of viscoelasticity.

Motivated by such considerations, Serrin [69] proposed a different version of the second law which does seem to deliver useful results more easily. In spirit, it is not so different from the first two versions and, for simpler kinds of theories, it leads to the same conclusions. Like them, it deals with cycles. Consider any thermodynamic system, letting $\mathbf{C}$ denote any cycle associated with it. Associate with this a function depending on a parameter Θ, $A(\mathbf{C}, \Theta)$ called the accumulation function, with the interpretation that

$$A(\mathbf{C}, \Theta) = \text{heat received by the system during the} \atop \text{cycle, at temperatures } \theta \leq \Theta. \tag{10.6.28}$$

Thus, for δQ interpreted as in (10.6.1), we have

$$\delta Q = A(\mathbf{C}, \infty). \tag{10.6.29}$$

In words, his version is

The accumulation function of a nonadiabatic cycle process *cannot be nonnegative.* $\tag{10.6.30}$

He explains how this relates to the Clausius and Kelvin–Planck versions, as he interprets them and, for this, I refer the reader to his paper. He shows that this statement implies that

$$\int_0^\infty \Theta^{-2} A(\mathbf{C}, \Theta) \, d\Theta \leq 0. \tag{10.6.31}$$

Under assumptions which are not very restrictive, one can put this in a form which looks more like the Clausius inequality. Suppose that, for $\mathbf{C}$ fixed, A is a sufficiently smooth function of Θ to enable us to integrate by parts. Suppose also that

$$\lim_{\Theta \to \infty} A/\Theta = \lim_{\Theta \to 0} A/\Theta = 0. \tag{10.6.32}$$

From (10.6.29), the first condition will certainly be satisfied if $\delta Q < \infty$, the second if no heat is received at temperatures near absolute zero and even less restrictive assumptions would suffice. With these assumptions, integration by parts gives, for any cycle,

$$\int_0^\infty \Theta^{-1} \, dA \leq 0. \tag{10.6.33}$$

Roughly, dA represents the heat received by the system between the temperatures Θ and $\Theta + d\Theta$, making this somewhat like the Clausius inequality.

Unlike the latter, there is no stipulation on how heat baths, etc., are to be used to effect heat transfer. This also resembles the rather mysterious integral mentioned by Gibbs, providing one interpretation which is meaningful.

Let us try evaluating (10.6.33) for a thermoelastic bar. For simplicity we will ignore heat conduction, so that

$$Q = \int_0^L r \, dx. \tag{10.6.34}$$

Consider a cycle beginning at $t = 0$ and ending at $t = T > 0$, described by functions $y(x,t)$ and $\theta(x,t)$, which should, of course, be defined for the whole bar, $0 \le x \le L$. The relevant points (x,t) in the x–t plane then form a rectangle R. For r, one can use Newton's law of cooling, the Stefan–Boltzmann law, or simply the function which satisfies (2.3.12) when some inputs are used to calculate the remaining terms. For simplicity, we assume that $r(x,t)$ and $\theta(x,t)$ are continuous functions on the rectangle, which implies that they are bounded. Also, we assume that $\theta(x,t)$ is bounded away from zero, so, for some positive constants a and b, we have

$$0 < a < \theta(x,t) \le b < \infty. \tag{10.6.35}$$

For convenience, we have picked a to be smaller than the minimum value of $\theta(x,t)$. Now divide up the temperature interval into N equal parts, $n = 1, \ldots, N$, giving intervals as indicated by

$$l_n : \Theta_n < \theta \le \Theta_{n+1}, \quad \Theta_1 = a, \quad \Theta_{N+1} = b, \\ \Theta_{n+1} - \Theta_n = (b - a)/N. \tag{10.6.36}$$

Now, we define subsets σ_n of the rectangle R by the condition that

$$(x,t) \in \sigma_n \Leftrightarrow \theta(x,t) \in l_n. \tag{10.6.37}$$

No point has two temperatures associated with it, so one point cannot belong to two such sets. Also, every point in R has a temperature associated with it, so the union of these disjoint sets is R,

$$\bigcup_{n=1}^N \sigma_n = R. \tag{10.6.38}$$

Now, from (10.6.28) and (10.6.36), for the cycle $\mathbf{C}$ considered,

$$A(\mathbf{C}, \Theta) = 0 \quad \text{if } \Theta \le a,$$

$$A(\mathbf{C}, \Theta) = \int_0^T \int_0^L r \, dx \, dt = \int_R r \, dx \, dt, \quad \text{if } \Theta \ge b, \tag{10.6.39}$$

which is consistent with (10.6.32). It is then clear that $dA = 0$, except when Θ is in the interval $a < \Theta \le b$, so the left side of (10.6.33) takes the form

$$\int_a^b \Theta^{-1} \, dA. \tag{10.6.40}$$

Now, from at least one definition of the integral we can approximate it as closely as we like in the manner indicated by

$$\int_a^b \Theta^{-1}\, dA \cong \sum_{n=1}^N \Theta_{n+1}^{-1} [A(\mathbf{C}, \Theta_{n+1}) - A(\mathbf{C}, \Theta_n)]. \qquad (10.6.41)$$

Now, the difference in the square brackets is, from the definition of A, the heat supplied at temperature θ in the range

$$\Theta_n < \theta \leq \Theta_{n+1}, \qquad (10.6.42)$$

that supplied at $\theta = \Theta_n$ being counted in both A's. From the definition of σ_n, this is given by

$$\int_{\sigma_n} r\, dx\, dt.$$

Thus, the above sum can be written as

$$\sum_{n=1}^N \Theta_{n+1}^{-1} \int_{\sigma_n} r\, dx\, dt. \qquad (10.6.43)$$

Now, using (10.6.38), we also have

$$\int_R (r/\theta)\, dx\, dt = \sum_{n=1}^N \int_{\sigma_n} (r/\theta)\, dx\, dt, \qquad (10.6.44)$$

and, with this, we can add and subtract this term, to put (10.6.41) in the form

$$\int_a^b \Theta^{-1}\, dA \cong \int_R (r/\theta)\, dx\, dt + \sum_{n=1}^N \int_{\sigma_n} r(\Theta_{n+1}^{-1} - \theta^{-1})\, dx\, dt. \qquad (10.6.45)$$

Note that, if N is large, θ is nearly equal to Θ_{n+1} in σ_n. By a more detailed analysis, which I will omit, this sum can be made arbitrarily small by taking N large enough. In the limit as $N \to \infty$, we obtain

$$\int_0^\infty \Theta^{-1}\, dA = \int_a^b \Theta^{-1}\, dA = \int_R (r/\theta)\, dx\, dt \leq 0, \qquad (10.6.46)$$

which is what we would get by applying the Clausius–Duhem inequality to a cycle when heat conduction is neglected ($q \equiv 0$). With a similar, but more tedious analysis, one can account for heat conduction which modifies (10.6.46) in the manner expected from analogous calculations based on the Clausius–Duhem inequality. Ignored here is the possibility that the sets σ_n may be so complicated that integrals over them are not well-defined. For a rigorous treatment, one needs analysis too complex to be used here, involving the theory of Lebesgue integration, but it is possible. In the three-dimensional case, when Q is given by (10.6.25), one can use similar arguments to show that, formally, (10.6.27) is equivalent to (10.6.33). For various kinds of theories of solids and fluids, the Clausius–Duhem inequality applies and gives sensible results. So this really provides numerous checks on the soundness of this version of the second law. With, say, plasticity theories, where one has doubts about the existence of entropy, it would be unsafe to assume that the Clausius–Duhem inequality applies. However, if you accept that (10.6.25) applies, you can use (10.6.27) as a mathematical representation of the second law, if you also accept this version. At least from my point of view, this version fits comfortably as part of the subject of classical thermodynamics, although it is relatively new. I prefer it to any of the three versions mentioned before.

There are also some different versions of the third law which deals with the unattainability of absolute zero ($\theta = 0$). As stated by Pippard [4], two of these are

By no finite series of processes is the absolute zero attainable,   (10.6.47)

and, what is perhaps better regarded as an addendum,

As the absolute temperature tends to zero, the magnitude of the entropy change in any reversible process tends to zero.   (10.6.48)

This provides support for the view that the second equality of (10.6.32) is not a very restrictive assumption, among other things. Various writers discuss this law and I will leave it to the interested reader to pursue this on his own.

Finally, some slightly different versions of another law, the zeroth law, are mentioned by various authors. As I interpret this, it is more concerned with the theory of temperature which I have not tried to discuss. In their treatment of this, Fosdick and Rajagopal [72] take as basic a version due to Maxwell,

(Maxwell) *Bodies whose temperatures are equal to that of the same body have themselves equal temperatures.*   (10.6.49)

In statements used by some others, the notion of equality of temperature is replaced by a statement that, pairwise, the bodies are in equilibrium with each other, a notion that I find misleading and confusing.

Newer than the theory of the old engines but certainly old enough to be regarded as classical, is another branch of thermodynamics, more related to statistical molecular theory. In this, Gibbs was also a pioneer. This involves quite different ideas about concepts of temperature, energy and entropy, more related to statistical averages. That these names are used here reflects the fact that some situations can be analysed with this and the older theory, with compatible results, if one agrees to use these names to make comparisons feasible. However, it does add confusion conceptually. It would take a very lengthy discussion to introduce these ideas and to compare them with those discussed above and I will not attempt this. Briefly, my experience is that each branch has its own virtues and faults. Pragmatically, I have found that one can be more successful than the other in dealing with a particular situation, so it helps to be familiar with both. For dealing with solids, one weak spot in this branch involves guesses which seem to need to be made to relate macroscopic deformation or motion to atomic or molecular motion. For crystals, Zanzotto [73] collects evidence that the commonly used hypothesis sometimes agrees and sometimes fails to agree with experiment. When it fails, we seem not to have a viable alternative. Workers interested in elastomers often use one hypothesis, although alternatives have been considered, to try to obtain some better agreement between molecular theory and macroscopic observations. At best, this is an indirect way of assessing the validity of the assumed relation. Suffice it to say there seems to be no panacea for problems of this kind.

I have mentioned only some of the difficulties encountered in trying to apply ideas of classical thermodynamics to solids. Of course, it is the difficulties which encourage intelligent workers to try to modify the old ideas in order to find better ways of treating the troublesome situations. I do not feel comfortable about giving general advice about how best to begin learning about such developments. Were I in this position, I would choose some questions of particular interest to me and ask experts for the best references relevant to these.

However, it is my experience that many workers are uneasy about some important concepts, particularly that of entropy. While it is not a panacea, I believe that it helps to understand how such concepts evolved, historically. Readers unfamiliar with this might find helpful the book by Truesdell [74], particularly his Historical Introit. Readers with much more mathematical ability than is required for my book might consult that by Šilhavy [75], which covers some topics touched upon here in greater depth, along with some omitted here.

References

[1] Truesdell, C., and Bharatha, S. (1977), *Classical Thermodynamics as a Theory of Heat Engines*, Springer-Verlag, Berlin.

[2] Kestin, J. (1966), *A Course in Thermodynamics*, Blaisdell, Waltham, MA.

[3] Tisza, L. (1966), *Generalized Thermodynamics*, M.I.T. Press, Cambridge, MA.

[4] Pippard, A.B. (1964), *The Elements of Classical Thermodynamics*, Cambridge University Press, Cambridge,.

[5] Bridgman, P.W. (1941), *The Nature of Thermodynamics*, Harvard University Press, Cambridge, MA.

[6] Bowen, R.M. (1989), *Introduction to Continuum Mechanics for Engineers*, Plenum Press, New York.

[7] Gibbs, G.W. (1875), On the equilibrium of heterogeneous substances. *Trans. Conn. Acad.*, **111**, 100 240, Gibbs, G.W. (1877-79), On the equilibrium of heterogeneous substances. *Trans. Conn. Acad.*, **111**, 343–524.

[8] Duhem, P. (1911), *Traité d'enérgétique ou thermodynamique générale* (two volumns), Gauthier-Villars, Paris.

[9] Einstein, A. (1956), *Investigations on the Theory of the Brownian Movement* (ed. R. Fürth; trans. A.D. Cowper), Dover Publications, New York.

[10] Pippard, A.B. (1985), *Response and Stability*, Cambridge University Press, Cambridge.

[11] Gelfand, I.M., and Fomin, S.V. (1963), *Calculus of Variations* (trans. R. Silverman), Prentice-Hall, Englewood Cliffs, NJ.

[12] Rivlin, R.S. (1975), The thermomechanics of materials with fading memory, in *Theoretical Rheology*, Applied Science Publishers Ltd, London.

[13] Bell, J.F. (1973), The experimental foundations of solid mechanics. *Handbuch der Physik*, vol. VIa/1, Springer-Verlag, Berlin.

[14] Kahl, G.D. (1967), Generalization of the Maxwell criterion for van der Waals' equation. *Phys. Rev.*, **155**, 78–80.

[15] Ericksen, J.L. (ed.) (1984), Proceedings of Workshop on Orienting Polymers, Minneapolis 1983. In *Lecture Notes in Mathematics*, vol. 1063, Springer-Verlag, Berlin.

[16] Landau, L.D. (1965), On the theory of phase transitions, in *Collected Papers of L.D. Landau* (ed. D. Ter Haar), Gordon and Breach and Pergamon Press, New York.

[17] Thomas, L.A., and Wooster, W.A. (1951), Piezocrescence: The growth of Dauphiné twinning in quartz under stress. *Proc. Roy. Soc. London*, **A208**, 43–62.

[18] Perkins, J. (ed.) (1975), *Shape Memory Effects in Alloys*, Plenum Press, New York.

[19] Delay, L., and Chandrasekaran, L. (eds.) (1982), Proc. Int. Conf. on Martensitic Transformations. *J. Physique*, **43**.

[20] Nishiyama, Z. (1978), *Martensitic Transformations*, Academic Press, New York.

[21] James, R.D., and Kinderlehrer, D. (1989), Theory of diffusionless phase transitions, in *Partial Differential Equations and Continuum Models of Phase Transitions* (eds. M. Rascle, D. Serre, and M. Slemrod), *Lecture Notes in Physics*, vol. 344, Springer-Verlag, New York.

[22] Sengers, A.L., Hocken, R., and Sengers, J.V. (1977), Critical point universality and fluids. *Phys. Today*, **30**, 42–51.

[23] Beatty, M.F. (1987), Topics of finite elasticity: hyperelasticity of rubber, elastomers, and biological tissues. *Appl. Mech. Rev.*, **40**, 1699–1734.

[24] Kitsche, W., Müller, I., and Strehlow, P. (1987), Simulation of pseudoelastic behavior in a system of balloons, in *Metastability and Incompletely Posed Problems* (eds. S. Antman, J.L. Ericksen, D. Kinderlehrer and I. Müller), *IMA Volumes in Mathematics and Its Applications*, vol. 3, Springer-Verlag, New York.

[25] Thompson, J.M.T., and Hunt, G.W. (1973), *A General Theory of Elastic Stability*, John Wiley and Sons, New York.

[26] Alexander, H. (1971), Tensile instability of initially spherical balloons. *Int. J. Eng. Sci.*, **9**, 151–162.

[27] Treloar, L.R.G. (1948), Stresses and birefringence in rubber subjected to general homogeneous strain. *Proc. Phys. Soc.*, **60**, 135–144.

[28] Kearsley, E.A. (1986), Asymmetric stretching of a symmetrically loaded elastic sheet. *Int. J. Solids Structures*, **22**, 111–119.

[29] Haughton, D.M., and Ogden, R.W. (1978), On the incremental equations in non-linear elasticity – II. Bifurcation of pressurized spherical shells. *J. Mech. Phys. Solids*, **26**, 111–138.

[30] Chen, Y.-C. (1987), Stability of homogeneous deformations of an incompressible elastic body under dead-load surface tractions. *J. Elasticity*, **17**, 223–248.

[31] Chen, Y.-C. (1988), Stability of pure homogeneous deformations of an elastic plate with fixed edges, *Q. J. Mech. Appl. Math.*, **41**, 249–264.

[32] Treloar, L.R.G. (1949), *The Physics of Rubber Elasticity*, Clarendon Press, Oxford.

[33] Courant, R., and Friedrichs, K.O. (1948), *Supersonic Flow and Shock Waves*, Interscience, New York.

[34] Dunn, J.E., and Fosdick, R.L. (1988), Steady, structured shock waves. Part 1: Thermoelastic materials. *Arch. Ration. Mech. Anal.*, **104**, 295–365.

[35] Gurtin, M.E. (1988), Toward a nonequilibrium thermodynamics of two-phase materials, *Arch. Ration. Mech. Anal.*, **100**, 275–312.

[36] Cheng, B.W. (1981), *Polymer Surfaces*, Cambridge University Press, Cambridge.

[37] Wu, S. (1982), *Polymer Interface and Adhesion*, Dekker, New York.

[38] Ricci, J.E. (1966), *The Phase Rule and Heterogeneous Equilibrium*, Dover, New York.

[39] Frazer, R.A., Duncan, W.J., and Collar, A.R. (1938), *Elementary Matrices*, Cambridge University Press, Cambridge.

[40] Gent, A.N., and Tompkins, D.A. (1969), Nucleation and growth of gas bubbles in elastomers. *J. Appl. Phys.* **40**, 2520–2525.

[41] Landau, L.D., Lifshitz, E.M., Kosevitch, A.M., and Pitaevskii, L.P. (1986), *Theory of Elasticity*, (3rd ed.) (trans. J.B. Sykes and W.H. Reid), Pergamon Press, Oxford.

[42] Frank, F.C. (1958), On the theory of liquid crystals, *Disc. Faraday Soc.*, **23**, 19–28.

[43] Courant, R., and Hilbert, D. (1953), *Methods of Mathematical Physics*, vol. 1, Interscience, New York.

[44] Chandrasekhar, S. (1977), *Liquid Crystals*, Cambridge University Press, Cambridge.

[45] de Gennes, P.G. (1974), *The Physics of Liquid Crystals*, Clarendon Press, Oxford.

[46] Virga, E.G. (1994), *Variational Theories of Liquid Crystals*, Chapman & Hall, London.

[47] Cohen, R.A. (1988), Fractional step methods for liquid crystal problems, Ph.D. Thesis, University of Minnesota, MN.

[48] Love, A.E.H. (1944), *A Treatise on the Mathematical Theory of Elasticity*, 4th ed., Dover, New York.

[49] Leipholz, H. (1970), *Stability Theory*, Academic Press, New York.

[50] Grindlay, J. (1970), *An Introduction to the Phenomenological Theory of Ferroelectricity*, Pergamon Press, Oxford.

[51] Brown, W.F., Jr. (1966), Magnetoelastic interactions, in *Springer Tracts in Natural Philosophy*, vol. 9 (ed. C. Truesdell), Springer-Verlag, New York.

[52] van der Waals, J.D. (1895), Théorie thermodynamique de la capillarité dans l'hypothèse d'une variation continué de densité. *Arch. Neerl. Sci. Exactes Nat.*, **28**, 121–209.

[53] Dunn, J.E., and Serrin, J. (1985), On the thermomechanics of interstitial working. *Arch. Ration. Mech. Anal.*, **88**, 95–133.

[54] Coleman, B.D., and Noll, W. (1963), The thermodynamics of elastic materials with heat conduction and viscosity. *Arch. Ration. Mech. Anal.*, **13**, 167–178.

[55] Müller, I. (1973), *Thermodynamik, Die Grundlagen der Material Theorie*, Bertelmann Universitätsverlag, Düsseldorf.

[56] Liu, I.-S. (1972), Method for exploitation of the entropy principle. *Arch. Ration. Mech. Anal.*, **46**, 131–148.

[57] Man, C.-S. (1985), Dynamic admissible states, negative absolute temperature and the entropy maximum principle. *Arch. Ration. Mech. Anal.*, **9**, 263–289.

[58] Poincaré, H. (1885), Sur l'équilibre d'une masse fluide animée d'un mouvement de rotation. *Acta Math.*, **7**, 259–380.

[59] Ericksen, J.L. (1992), Reversible and nondissipative processes. *Q. Jl. Mech. Appl. Math.*, **45**, 545–554.

[60] Meixner, J. (1973), Consistency of the Onsager–Casimir reciprocal relations. *Advance Mol. Relaxation Processes*, **5**, 319–331.

[61] Wollants, P., Roos, J.R., and Delaey, L. (1980), On the stress-dependence of the latent heat of transformations as related to the efficiency of a work performing cycle of a memory engine. *Scripta Met.*, **14**, 1217–1223.

[62] De Groot, S., and Mazur, P. (1962), *Nonequilibrium Thermodynamics*, North-Holland, Amsterdam.

[63] Green, A.E., and Naghdi, P.M. (1955), A general theory of an elastic-plastic continuum. *Arch. Ration. Mech. Anal.*, **18**, 251–281.

[64] Chaboche, J.L. (1988), Continuum damage mechanics. Parts I and II. *J. Appl. Mech.*, **55**, 59–72.

[65] Boltzmann, L. (1874), Zur Theorie der elastischen Nachwirkung, *Sitzber. Kaiserl. Akad. Wiss. Wien Math.-Naturw. Kl.*, **70**(II), 275–306.

[66] Ferry, J.D. (1970), *Viscoelastic Properties of Polymers* (2nd ed.), Wiley-Interscience, New York.

[67] Leitman, M.J., and Fisher, G.M.C. (1973), The linear theory of viscoelasticity, in *Handbuch der Physik*, vol. VIa/3 (ed. C. Truesdell), Springer-Verlag, New York, pp. 1–123.

[68] Coleman, B.D., and Owen, D.R. (1974), A mathematical foundation for thermodynamics. *Arch. Ration. Mech. Anal.*, **54**, 1–104.

[69] Serrin, J. (1979), Conceptual analyses of the classical laws of thermodynamics. *Arch. Ration. Mech. Anal.*, **70**, 355–371.

[70] Kestin, J. (1990), A note on the relation between the hypothesis of local equilibrium and the Clausius–Duhem inequality. *J. Non-Equilibrium Thermodynamics*, **15**, 193–212.

[71] Truesdell, C., and Toupin, R.A. (1960), The classical field theories, in *Handbuch det Physik*, vol. III/3. (ed. S. Flügge), 226–793, Springer-Verlag, New York.

[72] Fosdick, R.L., and Rajagopal, K.R. (1983), On the existence of a manifold for temperature. *Arch. Ration. Mech. Anal.*, **81**, 317–332.

[73] Zanzotto, G. (1988), Twinning in minerals and metals: remarks on the comparison of a thermoelastic theory, with some available experimental results, notes I and II. *Atti. Acc. Lincei Rend. Fis.*, **82**, 723–741 and 743–756.

[74] Truesdell, C. (1984), *Rational Thermodynamics* (2nd ed.), Springer-Verlag, New York.

[75] Šilhavy, M. (1997), *The Mechanics and Thermodynamics of Continuous Media*, Springer-Verlag, New York.

Index

Applied Mathematical Sciences

(continued from page ii)

(continued on next page)

Applied Mathematical Sciences

(continued from previous page)